To: MY Good

Don

Thanks for your support

Best Regards,

Rez

December 9, 1998

Fluid Catalytic Cracking Handbook

Fluid
Catalytic
Cracking
Handbook

Reza Sadeghbeigi

Gulf Publishing Company
Houston, Texas

Fluid Catalytic Cracking Handbook

Design, Operation, and Troubleshooting of FCC Facilities

Gulf Publishing Company
Book Division
P.O. Box 2608☐Houston, Texas 77252-2608

10   9   8   7   6   5   4   3

Library of Congress Cataloging-in-Publication Data
Sadeghbeigi, Reza.
    Fluid catalytic cracking handbook : design, operation, and troubleshooting of fcc facilities / Reza Sadeghbeigi.
      p.   cm.
    Includes bibliographical references (p.   –   ) and index.
    ISBN 0-88415-290-1
    1. Catalytic cracking.   I. Title.
TP690.4.S23   1995
665.5'33—dc20   95-24685
                                              CIP

Printed on Acid-Free Paper (∞).

To my wife Connie
and our children Jessica and Jason
for their understanding and support.

Contents

APPENDIX 10

Preface

Fluid Catalytic Cracking, better known as FCC, is one of the most important and complicated processes in petroleum refining. Since its first commercial start-up in 1942, thousands of articles have been written covering different aspects of this technology. However, my search of literature shows that there is no single book which discusses all the pertinent areas of FCC in a language that is clear, practical, and directed to the refinery process engineer.

My main objective is to share with readers the knowledge I have gained through 18 years of hands-on experience working in refineries and my continual analysis of cat cracking literature. This book is intended to provide readers with practical information covering all the key areas of FCC in a language that is clear to both the specialist and nonspecialist. The book will be used by engineers who are responsible for troubleshooting FCC units. It will be used by refinery management for the training of operating and technical personnel. It will become a key resource to research and development personnel, corporate planners, and companies servicing FCC facilities—that is, anyone having a vested interest in understanding and optimizing the FCC process.

For a process that is over 50 years old, one would think that FCC is a mature technology, but it is far from it. Cat cracking is an evolving technology that has been adapting to the changes in market and environmental demands. For instance:

- In the mid-1960s, there was a significant change in catalyst formulation that greatly improved product selectivity, resulting in higher gasoline yields and indirectly allowing refiners to process more feed in the unit.
- Further changes in the catalyst and hardware allowed processing of lower quality feeds in the FCC unit.
- With the governmental imposition of lead phase-down in motor gasoline in 1983, it was largely changes in catalyst, process, and hardware that resulted in meeting the octane requirements and

even exceeding them, as evidenced by the introduction of super-unleaded gasoline.

It is expected that FCC units will be playing an increasingly important role in meeting many challenges facing refiners. The passage of the Clean Air Act Amendment (CAAA) in 1990 affecting gasoline and diesel qualities and the implementation of Process Safety Management (PSM) are just two of the hot issues that the FCC technologist will be helping refinery management tackle in coming years.

One of the first things that new graduate refinery engineers recognize is that the refinery's operation can be more of an art than a science. This is particularly true when dealing with FCC. In my experience, if you put two cat cracker "experts" in a room, they will hardly ever agree. Just about all of the so-called FCC experts have some theories and hypotheses, but no one is totally sure what really takes place when the feed comes in contact with a 1300°F catalyst. One reason for the difficulty in understanding this process is that the interpretation of cause and effect is often subjective. Many improvements, such as feed and air injection devices, have been developed simply through trial-and-error approaches.

The goal of this book is to provide a straightforward yet thorough analysis of the fundamental issues affecting fluid catalytic cracking operations. The book would not have been possible without the support and feedback of my friends and colleagues. I express my sincere appreciation for their contributions.

Reza Sadeghbeigi

Fluid Catalytic Cracking

Handbook

CHAPTER 1

Process Flow
Description

Fluid catalytic cracking (FCC) is considered the primary conversion process in an integrated refinery. For many refiners, the cat cracker is the key to profitability in that the successful operation of the unit can determine whether or not a refiner can stay in business and remain competitive in today's market.

There are approximately 400 cat crackers operating worldwide, with a total processing capacity of over 12 million barrels per day [1]. Several oil companies, such as Exxon, Shell, and TOTAL, have their own designs; however, most of the current operating units have been designed or revamped by three engineering companies: UOP, M.W. Kellogg, and Stone & Webster. Although the mechanical configuration of individual FCC units may be arranged differently, their common objectives are to upgrade low-value feedstocks to more valuable products. It is important to note that, worldwide, about 45% of all gasoline produced comes from the FCC and ancillary units, such as the alkylation unit.

Since the start-up of the first commercial FCC unit in 1942, many improvements have been made to enhance the unit's mechanical reliability and its ability to crack heavier, lower-value feedstocks. FCC has a remarkable history of adapting to continual changes in market demands. Table 1-1 shows major developments in the cat cracking process.

The FCC unit utilizes a microspherodial catalyst which fluidizes when properly aerated. The main purpose of the unit is to convert high-boiling petroleum fractions called *gas oil* to high-value, high-octane gasoline and heating oil. Gas oil is the portion of crude oil that boils in the 650°–1050°+F (330°–550°C) range and contains a diversified mixture of paraffins, naphthenes, aromatics, and olefins (described in Chapter 2).

1

Table 1-1
The Evolution of FCC

1915	McAfee of Gulf Refining Co. discovered that a Friedel-Crafts aluminum chloride catalyst could catalytically crack heavy oil.
1936	Use of natural clays as catalyst greatly improved cracking efficiency.
1938	Standard of New Jersey, Kellogg, I.G. Farben, and Standard of Indiana formed a consortium to develop catalytic cracking.
1942	First commercial FCC unit (Model I) started up at Standard of New Jersey's Baton Rouge, La., refinery.
1947	First UOP stacked FCC unit was built. Kellogg introduced the Model III FCC unit.
1948	Davison Division of W.R. Grace & Co. developed microspherodial FCC catalyst.
1950s	Evolution of bed cracking process designs.
1956	Shell invented riser cracking.
1961	Kellogg and Phillips developed and put the first resid cracker onstream at Borger, Texas.
1964	Mobil Oil developed USY and ReY FCC catalysts.
1972	Amoco Oil invented high-temperature regeneration.
1974	Mobil Oil introduced CO promoter.
1975	Phillips Petroleum developed antimony for nickel passivation.
1981	TOTAL invented two-stage regeneration for processing residue.
1983	Mobil reported first commercial use of ZSM-5 octane/olefins additive in FCC.
1985	Mobil started installing closed cyclone systems in its FCC units.
1994	Coastal Corporation conducted commercial test of ultrashort residence time, selective cracking.

Contributor: Richard Wrench of M.W. Kellogg, December 1994.

Before proceeding, it is helpful to examine how a typical cat cracker fits into the refinery process. A petroleum refinery is composed of several processing units that are designed to convert raw crude oil into usable products such as gasoline, diesel, and jet fuel (Figure 1-1).

The crude unit is the first processing unit in the refining processes. Here, the raw crude is distilled into several intermediate products. The heavy portion of the crude oil that cannot be distilled in the atmospheric tower is heated and sent to the vacuum tower. The tar from

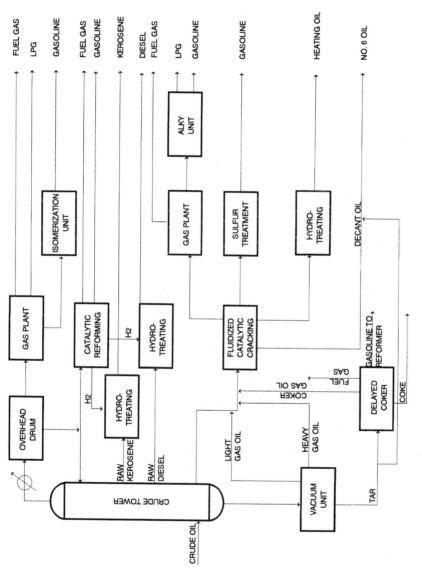

Figure 1-1. A typical high-conversion refinery.

the vacuum tower is sent to be processed further in a delayed coker, visbreaker, or other resid processing units.

The gas oil to a conventional cat cracker comes primarily from the atmospheric column, the vacuum tower, and the delayed coker unit. In addition, many refiners blend some atmospheric or vacuum residue with the cracker feedstocks to be processed in the FCC unit.

The FCC process is very complex. To provide a clear understanding of the unit operation, the process description has been broken down into six separate sections. These include the following:

- Feed Preheat
- Reactor
- Regenerator
- Main Fractionator
- Gas Plant
- Treating Facilities

1.1 FEED PREHEAT

Most refineries produce sufficient gas oil to meet the cat cracker's demand. However, in those refineries in which the gas oil produced does not meet the cat cracker capacity, it may be economical to supplement feed by purchasing FCC feedstocks or blending some residue. The refinery-produced gas oil and any supplemental feedstocks are generally combined and sent to a surge drum which provides a steady flow of feed to the FCC unit's charge pumps. This drum can also serve as a device to separate any water or vapor that may be inherent in the feedstocks.

From the surge drum, the feed is normally heated to a temperature of 550°–700°F (270°–357°C). The main fractionator bottoms pump-around and/or fired heaters are usually the sources of heat to preheat the feed. The feed preheater provides a tool to easily vary catalyst-to-oil ratio. In units where the air blower is the constraint, increasing preheat temperature allows increased throughput. The effects of feed preheat are discussed in Chapter 6.

1.2 REACTOR

The reactor-regenerator (Figure 1-2) is the heart of the FCC process. In a modern cat cracker, virtually all the reactions occur in the riser over a short period of two to four seconds before the catalyst and the

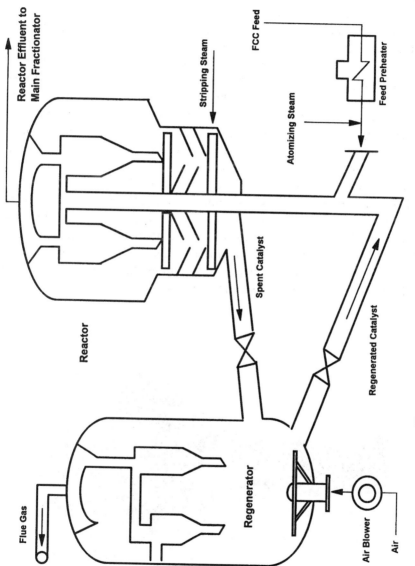

Figure 1-2. A typical FCC reactor-regenerator.

products are separated in the reactor. However, some thermal and non-selective catalytic cracking reactions continue to occur in the reactor housing. A number of refineries are modifying the riser termination devices to eliminate this problem.

From the preheater, the feed enters the riser near the base where it contacts the incoming regenerated catalyst. The ratio of catalyst to oil is normally in the range of 4:1 to 9:1 by weight. The heat absorbed by the catalyst in the regenerator provides the energy to heat the feed to desired reactor temperature. The net heat of the reaction occurring in the riser is enthothermic, i.e., it requires energy input. This energy is provided by the circulating catalyst. The catalytic reactions occur in the vapor phase; as soon as the feed is vaporized, cracking reactions begin instantaneously. The expanding volume of the vapors that are generated will lift the catalyst and carry it up the riser.

The riser (Figure 1-3) is essentially a vertical pipe usually having a 4- to 5-inch-thick refractory lining for insulation and abrasion resistance. Typical riser dimensions are 2 to 6 feet in diameter and 75 to 120 feet in length. The ideal riser simulates a plug flow reactor, where the catalyst and the vapor travel along the length of the riser at the same velocity with minimum back-mixing.

Figure 1-3. A typical FCC riser (60" diameter). *(Courtesy of VAL-VAMP, Incorporated, Houston, Texas.)*

Efficient contacting of the feed and the catalyst is critical for achieving the desired cracking reactions. Steam is commonly used to atomize the feed because feed atomization increases the availability of feed at the reactive acid sites on the catalyst. With employment of a high-activity zeolite catalyst, virtually all of the cracking reactions take place in the riser in a less than two-second time frame. Risers are normally designed for an outlet vapor velocity of 50 ft/sec to 75 ft/sec, with an average hydrocarbon residence time of about two seconds (based on outlet conditions). As a consequence of the cracking reactions, a hydrogen-deficient material called *coke* is deposited on the catalyst, reducing catalyst activity.

1.2.1 Catalyst Separation

After exiting the riser, the catalyst enters the reactor. In today's FCC operations, the reactor serves basically two functions: as a disengaging space for the separation of catalyst and vapor, and as the reactor cyclone's housing.

Nearly every FCC unit employs some type of inertial separation device connected on the end of the riser to separate the bulk of the catalyst from the vapors. Most units use a deflector device to turn catalyst direction downward. On some units, the riser is directly attached to a set of cyclones. The term "rough cut" cyclones (Figure 1-4) generally refers to this type of arrangement. These schemes separate approximately 75%–99% of the catalyst from the product vapors.

Most FCC units employ either single or two-stage cyclones (Figure 1-5) to separate the remaining catalyst particles from the cracked vapors. The cyclones collect the catalyst and return it to the stripper through the use of diplegs and flapper valves. The product vapors exit the cyclones and flow to the main fractionator column for recovery. The efficiency of a typical two-stage cyclone system is 99.995+%.

It is important to incorporate process and mechanical provisions to separate catalyst and vapors as soon as they enter the reactor, otherwise, the extended contact of the vapors with the catalyst in the reactor will allow recracking of some of the desirable products. Furthermore, the extended residence time also promotes thermal cracking of the desirable products.

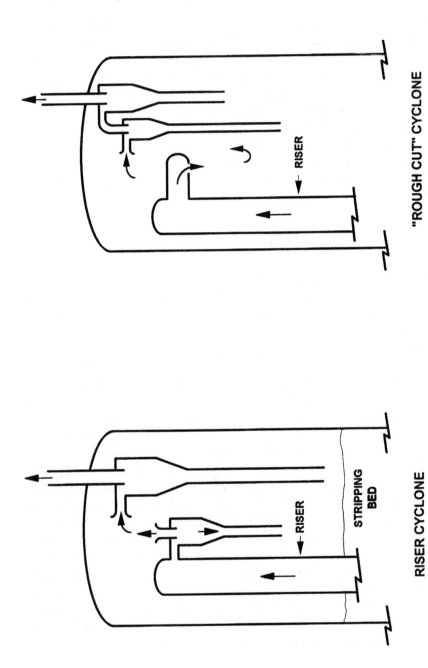

RISER CYCLONE

"ROUGH CUT" CYCLONE

Figure 1-4. Examples of riser and "rough cut" cyclones.

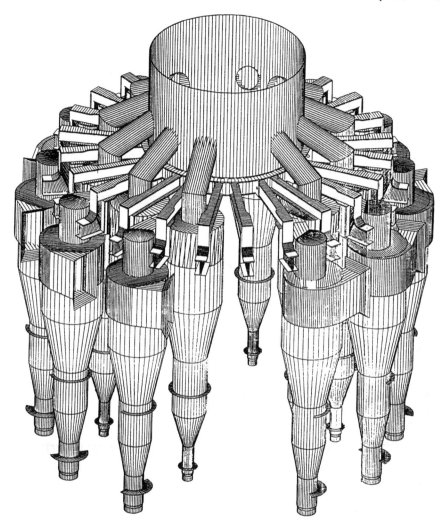

Figure 1-5. A two-stage cyclone system. *(Courtesy of Bill Dougherty, BP Oil Refinery, Marcus Hook, Pa.)*

1.2.2 Stripping Section

As the spent catalyst falls into the stripper, valuable hydrocarbons are adsorbed within the catalyst bed. Stripping steam, at a rate of 2–5 lbs per 1,000 lbs of circulating catalyst, is used to strip these hydrocarbons from the catalyst. Both baffled and unbaffled stripper designs (Figure 1-6) are in commercial use. An efficient stripper design will incorporate optimum countercurrent contacting between the catalyst and steam. Reactor strippers are commonly designed for a steam superficial velocity of 0.75 ft/sec and a catalyst flux rate of 500 to 700 lbs of circulating catalyst per minute per square foot. At too high a flux, the falling catalyst tends to entrain steam, thus reducing the effectiveness of stripping steam.

It should be noted that not all the hydrocarbon vapors can be displaced from the catalyst pores in the stripper. A fraction of them are

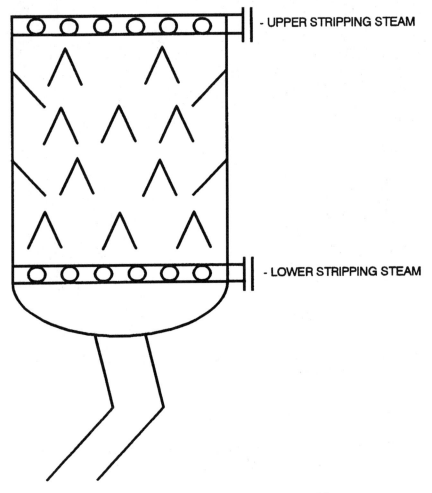

Figure 1-6.   An example of a two-stage stripper.

carried with the spent catalyst into the regenerator. These vapors have a higher hydrogen to carbon ratio than the coke on the catalyst. The drawbacks of allowing these hydrogen-rich hydrocarbons to enter the regenerator are as follows:

- Loss of liquid product. Instead of the hydrocarbons burning in the regenerator, they could be recovered as liquid products.
- Loss of throughput. The combustion of hydrogen to water produces 3.7 times more heat than the combustion of carbon to carbon dioxide. The increase in the regenerator temperature caused by excess hydrocarbons could exceed the temperature limit of the regenerator internals and force the unit into a reduced feed rate mode of operation.
- Loss of catalyst activity. The higher regenerator temperature combined with formation of steam in the regenerator reduces catalyst activity by destroying the catalyst's crystalline structure.

The flow of the spent catalyst to the regenerator is typically controlled by the use of a valve that slides back and forth. This slide valve (Figure 1-7) is used to control the catalyst level in the stripper. The catalyst level in the stripper provides the pressure head which allows the catalyst to flow into the regenerator. The exposed surface of the slide valve is usually lined with a suitable refractory to withstand erosion.

1.3 REGENERATOR

The regenerator has two main functions: It restores catalyst activity and supplies heat to crack the feed. The spent catalyst entering the regenerator contains between 0.8 and 2.5 wt% coke, depending on the quality of the feedstocks. Components of coke are carbon, hydrogen, and trace amounts of sulfur and nitrogen. These burn according to the following reactions:

			K Cal/Kg of C, H_2, or S	BTU/lb of C, H_2, or S	
$C + 1/2 \; O_2$	$\rightarrow$	CO	2,200	3,968	(1-1)
$2CO + O_2$	$\rightarrow$	$2CO_2$	5,600	10,100	(1-2)
$C + O_2$	$\rightarrow$	CO_2	7,820	14,100	(1-3)
$H_2 + 1/2 \; O_2$	$\rightarrow$	H_2O	28,900	52,125	(1-4)
$S + xO$	$\rightarrow$	SO_x	2,209	3,983	(1-5)
$N + xO$	$\rightarrow$	NO_x			(1-6)

Figure 1-7. Catalyst slide valve. *(Courtesy of Enpro Systems, Channelview, Texas.)*

Air is the source of oxygen for the combustion of coke and is supplied by a large air blower. The air blower provides sufficient air velocity and pressure to maintain the catalyst bed in a fluid state. The air enters the regenerator through an air distributor (Figure 1-8) located near the bottom of the vessel. The design of an air distributor is important in achieving efficient and reliable catalyst regeneration, and a number of designs are available from the major FCC unit licensers. Generally, air distributors are designed for a 1.0–2.0 psi pressure drop to ensure positive air flow through all nozzles.

There are two regions in the regenerator: the *dense phase* and the *dilute phase*. At the velocities common in the regenerator, 2–4 ft/sec, the bulk of catalyst particles are located in the dense bed immediately above the air distributor. The dilute phase is the region above the dense phase up to the cyclone inlet, and has a substantially lower catalyst concentration.

1.3.1 Standpipe/Slide Valve

From the regenerator, the regenerated catalyst flows down a transfer line commonly referred to as a standpipe. A standpipe provides the necessary pressure head needed to circulate the catalyst around the unit. In some units, standpipes are extended into the regenerator, and the top section is often called a catalyst hopper. The hopper, being internal to the regenerator, is usually of an inverted cone design. The function of the hopper is to provide a sufficient time for the regenerated catalyst to be aerated before entering the standpipe. Standpipes are typically sized for a flux rate in the range of 100 lbs/sec.ft^2 to 300 lbs/sec.ft^2 of circulating catalyst. In most cases, sufficient flue gas is carried down with the regenerated catalyst to keep it fluidized. However, longer standpipes may require external aeration to ensure that the catalyst remains fluidized. Aeration is ensured by injecting a supplemental gas medium, such as air, steam, nitrogen, or fuel gas, along the length of the standpipe. The catalyst density in a well-designed standpipe is in the range of 35 lbs/ft^3 to 45 lbs/ft^3.

The flow rate of the regenerated catalyst to the riser is commonly regulated through the use of either a slide or plug valve. The operation of a slide valve is similar to that of a variable orifice. Slide valve operation is automatic. Its main function is to supply enough catalyst to heat the feed and achieve the desired reactor temperature.

PIPE GRID DESIGN

AIR RING DESIGN

Figure 1-8. Examples of air distributors. *(Top: courtesy of Enpro Systems, Inc., Channelview, Texas; bottom: courtesy of VAL-VAMP, Incorporated, Houston, Texas.)*

1.3.2 Catalyst Separation

As the flue gas leaves the dense phase of the regenerator, it entrains catalyst particles. The amount of entrainment depends largely on the flue gas superficial velocity. The larger catalyst particles, 50–90 μ, fall back into the dense bed. However, for the smaller particles, 0–50 μ, the flue gas velocity is sufficient to suspend them in the dilute phase and carry them out of the regenerator and into the cyclones.

Most FCC unit regenerators employ 6 to 16 sets of primary and secondary cyclones in series, depending on unit size. These cyclones are designed to recover catalyst particles that are greater than 20 microns. The recovered catalyst particles are returned to the regenerator via the diplegs. The flue gas exits the cyclones through a plenum chamber.

The distance above the catalyst bed in which the flue gas velocity has stabilized is referred to as the transport disengaging height (TDH). At this distance, there is no further gravitation of catalyst. The centerline of the first-stage cyclone inlets should be at TDH or higher; otherwise, excessive catalyst entrainment will cause extreme catalyst losses.

1.3.3 Flue Gas Heat Recovery Schemes

The hot flue gas leaving the regenerator plenum holds an appreciable amount of energy. A number of heat recovery schemes are used to recover this energy. In some units, the flue gas is sent to a CO boiler where both the sensible and the combustible heat are used to generate high-pressure steam. In other units, the flue gas is exchanged with boiler feed water to produce steam via the use of shell/tube or box heat exchanger.

In most units, the flue gas pressure is let down to atmospheric pressure across an orifice chamber. The orifice chamber is a vessel containing a series of perforated plates designed to maintain a given back pressure upstream of the regenerator pressure control valve.

In some larger units, a turbo expander is used to recover this pressure energy. To protect the expander blades from being eroded by the catalyst, the flue gas is first sent to a third-stage separator to remove the fines. The third-stage separator, which is external to the regenerator, consists of a large number of swirl tubes designed to separate 70% to 95% of the incoming particles from the flue gas.

A power recovery train (Figure 1-9) employing a turbo expander usually consists of four parts: the expander, a motor/generator, an air blower, and a steam turbine. The steam turbine is used primarily for start-up and often to supplement the expander for generation of electricity.

The motor/generator can produce or use power. In some FCC units, the expander horsepower exceeds the power needed to drive the air blower. In this case, the excess power is transmitted to the refinery electrical system. If the expander generates less power than what is required by the blower, the motor/generator provides the power to hold the power train at the desired speed.

From the expander, the flue gas goes through a steam generator to recover additional energy. Depending on local environmental regulations, an electrostatic precipitator (ESP) or a wet gas scrubber may be placed downstream of the waste heat generator prior to release of the flue gas to the atmosphere. In some units, an ESP is used to further remove catalyst fines in the range of 5–20 μ from the flue gas. In other units, a wet gas scrubber is employed to remove both catalyst fines and sulfur compounds from the flue gas stream.

1.3.4 Partial versus Complete Combustion

There are two methods of regeneration practiced by FCC units: *partial combustion* and *complete combustion*. A number of cat crackers that process heavy residue feedstocks use a two-stage regeneration. The methods of regeneration are differentiated from one another by the operating temperature and efficiency of carbon removal from the catalyst particles.

In the partial combustion mode, some of the carbon on the spent catalyst is incompletely burned to carbon monoxide. As shown in Section 1.3, the oxidation CO to CO_2 generates 2.5 times more heat than combustion of C to CO. Refiners operate regenerators in partial burn so as not to exceed the temperature rating of the regenerator internals. In addition, switching to a partial burn mode can allow cat crackers to process heavier feedstocks and still remain within regenerator temperature limits. In the partial combustion mode, the regenerator temperature is controlled by regulating flow of the combustion air.

Full combustion, or high-temperature, regeneration was introduced in 1972. Until the early 1970s, the design temperature of most regenerator cyclone supports was limited to around 1250°F. The evolution

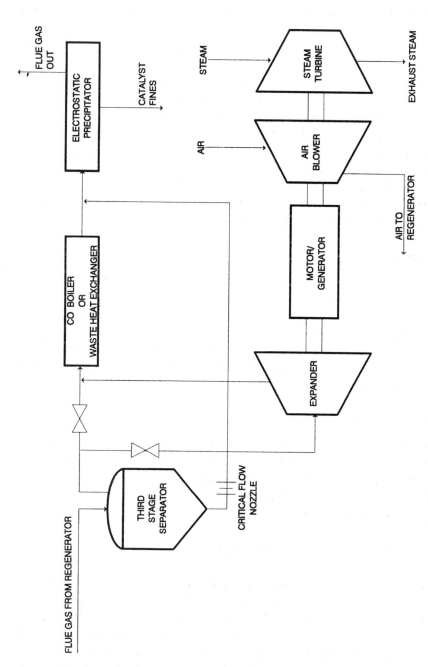

Figure 1-9. A typical flue gas power recovery scheme.

of complete combustion and its subsequent yield improvements came about accidentally.

The full combustion mode of regeneration utilizes excess oxygen for complete CO combustion and reduction of carbon on the regenerated catalyst to less than 0.10 wt%. The reduced level of carbon on the regenerated catalyst increases the catalyst's activity and selectivity. Full combustion can be achieved either thermally or with the aid of a combustion promoter. Thermal conversion of CO to CO_2 can occur in a properly designed regenerator; however, because ideal conditions sometimes cannot be reached, most cat crackers use a CO combustion promoter catalyst to encourage CO combustion in the dense phase. The promoter contains low levels of platinum and/or palladium. Full combustion has both advantages and disadvantages.

Advantages

- Higher catalyst selectivity because of lower carbon on the regenerated catalyst.
- Lower CO content of flue gas, allowing it to be released to the atmosphere.
- Higher heat release in the regenerator which will be beneficial when processing hydrotreated stocks that do not produce sufficient coke for heat balance.

Disadvantages

- Lower cat/oil ratio because of higher regenerator temperature.
- Harder impact on catalyst activity due to thermal deactivation.
- Greater requirement for combustion air.
- More expensive metallurgy in the regenerator to accommodate higher temperature operation.

1.3.5 Catalyst Handling Facilities

Even with proper operation of the reactor and regenerator cyclones, catalyst particles smaller than 20 microns still escape from both of these vessels. The catalyst fines escaping the reactor collect in the fractionator bottoms product storage tank. The recoverable catalyst fines exiting the regenerator are removed by the electrostatic precipitator. The catalyst losses are related mainly to hydrocarbon vapor and flue gas velocities, the catalyst's physical properties, and attrition due to collision of catalyst particles with the vessels' internals, and other catalyst particles.

The activity of the catalyst degrades with time. The loss of activity is attributed primarily to the impurities in the FCC feed, such as nickel, vanadium, and sulfur, and to thermal and hydrothermal deactivation mechanisms. To maintain the desired activity, fresh catalyst is continually added to the unit. Fresh catalyst is stored in a fresh catalyst hopper and, in most units, is added automatically to the regenerator via a catalyst loader.

The circulating catalyst in the FCC unit is called *equilibrium catalyst,* or simply E-cat. Periodically, quantities of equilibrium catalyst are withdrawn and stored in the E-cat hopper for future disposal. A refinery that processes residue feedstocks in its cat cracker can make use of a good-quality E-cat from another refinery that processes light sweet feed. Residue feedstocks contain large quantities of impurities such as metals. Therefore, the use of a good-quality E-cat in conjunction with fresh catalyst can be cost-effective in maintaining low catalyst costs.

1.4 MAIN FRACTIONATOR

The purpose of the main fractionator (Figure 1-10) is to desuperheat and recover liquid products from the reactor vapor. The hot-product vapors from the reactor flow into the main fractionator; these vapors enter the column near the base. The main function of the fractionator is to condense and separate the reaction products. The fractionation is accomplished by condensing and revaporizing hydrocarbon components as the vapor flows upward through trays in the tower.

The operation of the main column is similar to a crude tower but with two differences. First, the effluent vapors must be cooled before any fractionation begins. Second, large quantities of gases will go overhead with the unstabilized gasoline for further separation.

The main purpose of the bottom section of the main column is to provide a heat transfer zone. Shed decks, disk/doughnut trays, and grid packing are among some of the devices used to desuperheat and cool the overhead vapor. The cooled pumparound also serves as a scrubbing medium to wash down catalyst fines entrained in the vapors. The flow of quench is adjusted to maintain the fractionator bottoms temperature

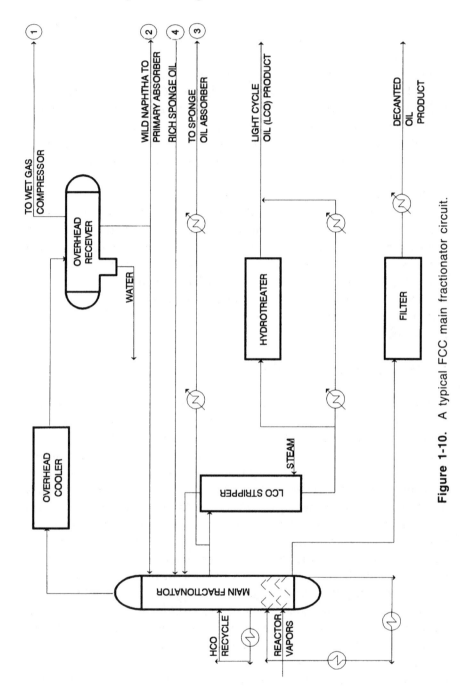

Figure 1-10. A typical FCC main fractionator circuit.

below the value that causes coking. Most FCC operators maintain this temperature below 700°F (370°C).

The recovered heat from the main column bottoms is used to preheat the fresh feed, generate steam, serve as a heating medium for the gas plant reboilers, or some combination of these streams. The heaviest product from the main column is commonly called slurry or decant oil. In this book, these terms are used interchangeably. The decant oil is often used as a "cutter stock" to make No. 6 fuel oil. The quality of some decant oil is such that it can be used for higher-value carbon black feedstocks.

Earlier FCC units used high recycle rates, slurry settlers, and other centrifugal separators to remove catalyst fines from decanted oil. A slipstream of FCC feed is employed as a carrier to return the collected fines from the bottom of the separator to the riser. However, improvements in the physical properties of FCC catalyst and in the reactor cyclones have lowered the catalyst carry-over to the point that many units operate without separators. Instead, the decanted oil is sent directly to the storage tank. In time, catalyst fines will accumulate in these tanks and require some form of disposal. Nevertheless, a number of units continue to use some form of slurry settlers to minimize ash content of the decanted oil.

Aside from decanted oil product, the main column is often designed to have three possible sidecuts: heavy cycle oil, light cycle oil, and heavy naphtha. In many units, the light cycle oil (LCO) is the only sidecut that leaves the unit as a product. LCO is withdrawn from the main column and routed to a side stripper. After being stripped by steam for flash control, LCO is pumped to storage or to a diesel hydrotreater for sulfur removal prior to being blended into the heating oil pool. In some units, a slipstream of the stripped LCO, called "sponge oil," is sent to the sponge oil absorber at the gas plant. In other units, sponge oil is the cooled, unstripped LCO.

The heavy cycle oil, heavy naphtha, and other circulating side reflux streams are used to remove heat from the fractionator and to supply heat for the unsaturated gas recovery section. The amount of heat removed at any pumparound point is set to evenly distribute vapor and liquid loads throughout the column and to provide the necessary internal reflux.

Unstabilized gasoline and light gases pass up through the main column and leave as vapor. The overhead vapor is cooled and partially condensed in the fractionator overhead condensers.

1.5 GAS PLANT

The role of the FCC gas plant (Figure 1-11) is to separate the unstabilized gasoline and light gases into fuel gas, C_3 and C_4 compounds, and gasoline. C_3's and C_4's include propane, propylene, normal butane, isobutane, and butylene. The propylene and butylene are generally alkylated in a sulfuric or hydrofluoric alkylation unit to produce high-octane gasoline. Additionally, most gas plants also include treating facilities to remove sulfur from these products.

The partially condensed overhead stream from the main fractionator overhead condensers contains vapors, hydrocarbon liquid, and water. They flow to an overhead receiver operating at a low pressure (<15 psig). In the overhead accumulator drum, hydrocarbon vapor, hydrocarbon liquid, and water are separated.

The hydrocarbon vapors flow to a *wet gas* compressor. At the operating conditions of the overhead receiver, the gas stream to the compressor contains not only ethane and lighters but about 90% C_3's and C_4's as well as about 10% produced gasoline. The phrase "wet gas" refers to condensable components of the gas stream.

An external two-stage centrifugal compressor is typically employed to raise the pressure of the gas stream. This type of compressor generally incorporates multistage turbines that are driven by high-pressure steam. The steam is exhausted to a surface condenser operating under vacuum.

The vapors from the compressor's first-stage discharge are partially condensed and flashed in an interstage drum. The liquid hydrocarbon is pumped either to a high-pressure separator (HPS) or directly to the stripper.

The vapor from the interstage drum flows to a second-stage compressor. From the second-stage discharge, the compressed vapor is often mixed with gases and LPG streams from other refinery units. Examples of these streams are: the primary stripper overhead vapor, the primary absorber rich oil, and wash water. This mixture is partially condensed and flashed in the HPS. Wash water is injected mainly to dilute contaminants, such as ammonium salts, that can cause equipment fouling.

The vapor and liquid from the HPS flow to the primary absorber and stripper, respectively. The HPS is essentially a separation stage with an external cooler located between the primary stripper and absorber.

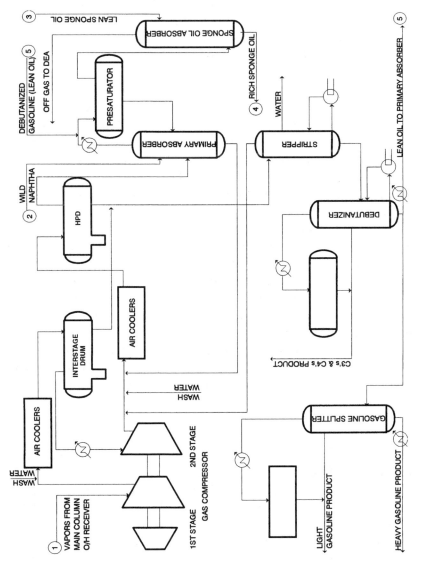

Figure 1-11. A typical FCC gas plant.

1.5.1 Primary Absorber

The purpose of the primary absorber is to recover C_3's and heavier components from the HPS overhead vapors. The HPS overhead vapor contains appreciable amounts of C_3's and heavier components. To recover these components, the HPS vapor flows to an absorber, entering below the bottom tray. Frequently, there are two sources of absorption oils that are utilized jointly in this tower. The first one is the hydrocarbon liquid from the main fractionator overhead receiver. This stream, often called "wild," or unstabilized, naphtha, enters the absorber a few trays below the top tray. The second absorbent is the cooled debutanized gasoline. The expression "lean oil" generally refers to the debutanized gasoline plus the unstabilized naphtha from the overhead receiver.

To enhance C_3+ recovery, a number of FCC gas plants have installed presaturator drums that function as an additional absorption stage. In this operation, the cooled-debutanized gasoline is first mixed (presaturated) with the absorber overhead gas. The mixture is cooled and flashed in the presaturator drum. The liquid from this drum is then pumped to the top of the primary absorber.

The gas from the presaturator drum is sent to a sponge oil absorber. The contacting of debutanized gasoline and primary absorber gas in the presaturator drum results in C_5+ pickup into the gas. This is removed in the sponge oil absorber.

The absorption process is exothermic. Therefore, to improve C_3+ recovery, liquid collected in one of the middle trays is withdrawn and pumped through an intercooler and returned to the tray below.

1.5.2 Sponge Oil Absorber

The vapor from the primary absorber or the presaturator contains a small quantity of gasoline. The purpose of a sponge oil absorber is to recover this gasoline. "Sponge oil" refers to stripped or unstripped light cycle oil and is employed to provide the final absorption of the dry gas stream. Instead of LCO, a few FCC units use cooled-heavy naphtha as a sponge oil. The heavy naphtha is drawn from the main fractionator; however, to recover the same amount of gasoline, more heavy naphtha must be circulated.

The water-cooled sponge oil enters the absorber on the top tray. The gas from the presaturator or from the primary absorber (in the absence of a presaturator) flows into the sponge oil absorber on the bottom tray. The rich sponge oil is then returned to the main fractionator. The lean gas leaves the absorber to a sweetening unit prior to entering the refinery fuel gas system.

1.5.3 Stripper or De-ethanizer

The HPS liquid consists mostly of C_3's and heavier hydrocarbons; however, it also contains small fractions of C_2's, H_2S, and entrained water. The purpose of the stripper is to remove these light-ends. The liquid enters the stripper on the top tray. The required heat for stripping is provided by an external reboiler. Examples of heating media are steam and the debutanizer bottoms. The vapor that is generated in the reboiler rises through the tower and strips the lighter fractions from the descending liquid. The rich vapor then flows to the HPS via the condenser. The stripped naphtha leaves the tower bottoms and goes to the debutanizer. Usually, there is at least one water draw in the tower to remove the entrained water.

1.5.4 Debutanizer

The stripper bottoms contain C_3's, C_4's, and gasoline, and the debutanizer is designed to separate the C_3's and C_4's from the gasoline. The stripper bottoms enter the debutanizer about midway in the tower. In some units, the feed is preheated before entering the debutanizer. In most other units, the stripper bottoms are sent directly to the debutanizer. The operating pressure of the debutanizer is approximately 50 psi lower than the stripper. A control valve that regulates stripper bottoms level is the means of this pressure drop. As a result of this drop, part of the feed is vaporized.

The debutanizer separates the feed into two products. The overhead product contains a mixture of C_3's and C_4's. The bottoms product is the stabilized gasoline. Heat for separating these products comes from an external reboiler. The heating source is usually the main fractionator bottoms pumparound and/or steam.

The overhead product is totally liquefied in the overhead condensers. A portion of the overhead liquid is pumped and returned to the tower as reflux. The remainder is sent to a sweetening unit to remove H_2S and other organic sulfur compounds. From the sweetening unit, the mixed C_3's and C_4's are typically fed to an alkylation unit. In some refineries, the overhead is pumped to a depropanizer tower where the C_3's are separated from C_4's. The C_3's are processed as petrochemical feedstocks and the C_4's are alkylated.

The debutanized gasoline is cooled first by supplying heat to the stripper reboiler or preheating the debutanizer feed. This is followed by a set of air or water coolers. A portion of the debutanizer bottoms is pumped as a lean oil to the presaturator or, in the absence of a presaturator, to a primary absorber. The balance is treated for sulfur removal and blended into the refinery gasoline pool.

1.5.5 Gasoline Splitter

A number of refiners split the debutanized gasoline into "light" and "heavy" gasolines. The splitting is done to optimize the refinery gasoline blending pool. In a few gasoline splitters, a third "heart cut" is withdrawn. This intermediate cut is low in octane, and it is processed in another unit for further upgrading of its octane.

1.5.6 Water Wash System

To retard corrosion and hydrogen blistering attacks in the gas plant equipment, water is injected into the wet-gas compressor discharge. The water reduces corrosion potential and keeps salts in solution.

The cat cracker feedstock contains low concentrations of organic sulfur and nitrogen compounds. Cracking of organic nitrogen compounds liberates cyanide (HCN), ammonia (NH_3), and other nitrogen compounds. Cracking of organic sulfur compounds produces hydrogen sulfide (H_2S) and other sulfur compounds.

Water comes from condensation of process steam in the main fractionator overhead condensers. In the presence of H_2S, NH_3, and HCN, a wet environment exists in the FCC gas plant. This environment is conducive to corrosion attacks.

The chemical reactions are as follows:

$$H_2S \rightarrow 2H^- + 2HS^- \tag{1-7}$$

$$Fe + 2HS^- \rightarrow FeS + S^{-2} + 2H° \tag{1-8}$$

$$FeS + 6CN^- \rightarrow Fe(CN)_6^{-4} + S^{-2} \tag{1-9}$$

Without the presence of cyanide, the iron sulfide scale can provide a layer of protection to the process equipment. This scale actually inhibits further corrosion attack. However, as shown in equation 1-9 above, the cyanide removes this protective scale and exposes fresh iron to further corrosion attacks. Additionally, without cyanide, the atomic $H°$ combines to form gaseous H_2 which is too large to diffuse through steel [2]. The presence of cyanide allows penetration of $H°$ into the steel, causing hydrogen blistering.

Most refiners employ a continuous water wash as the principal method of retarding corrosion and hydrogen blistering. The best source of water is either a steam condensate or a well-stripped water from a sour water stripper.

In the cat cracker gas plant, the greatest concentration of corrosion agents (H_2S, HCN, and NH_3) is found at high-pressure points. Water is usually injected into the first- and second-stage compressor discharges. The water contacts the hot gas and scrubs these agents. There are two common injection methods: *forward cascading* and *reverse cascading*. In forward cascading (Figure 1-12A), the water is normally injected into the discharge of the first-stage compressor. From the interstage drum, the water is pumped to the second stage discharge. From the high pressure separator, the water is then pressured to the sour water stripper.

In reverse cascading (Figure 1-12B), fresh water is injected into the second-stage discharge. The water containing corrosion agents is pressured to the first-stage discharge and then on to the main fractionator overhead. From the overhead receiver, the water is then pumped to the sour water stripper. Reverse cascading requires one less pump, but a portion of cyanide captured in the second discharge is released in the interstage, thus forming a cyanide recycle. Consequently, forward cascading is more effective in minimizing cyanide attack.

1.6 TREATING FACILITIES

The gas plant products, namely fuel gas, C_3's, C_4's, and gasoline, contain varying levels of sulfur compounds that require treatment. Impurities in the gas plant products are acidic in nature. Examples of these impurities are hydrogen sulfide (H_2S), carbon dioxide (CO_2),

(text continued on page 30)

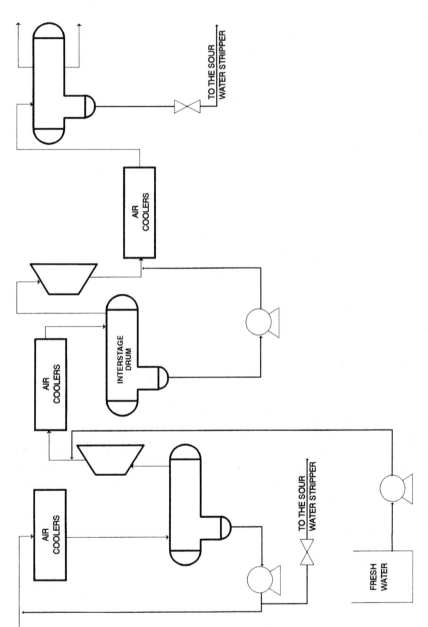

Figure 1-12A. A typical *forward* cascading scheme for water wash.

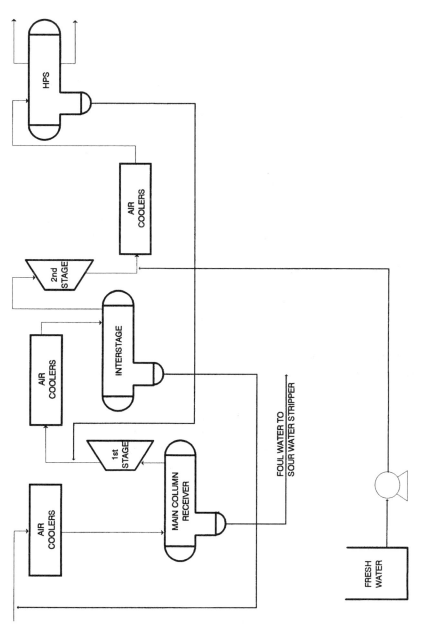

Figure 1-12B. A typical *reverse* cascading scheme for water wash.

(text continued from page 27)

mercaptan (R-SH), phenol (ArOH), and naphthenic acids (R-COOH). Other impurities such as carbonyl and elemental sulfur may also be present in the above streams.

Refineries use a family of amine and caustic solutions to remove these impurities. The amine solvents known as alkanolamines remove both H_2S and CO_2. Hydrogen sulfide is poisonous and toxic. For the gas used in the refinery furnaces and boilers, the maximum H_2S concentration is normally about 160 ppm.

Amines remove the bulk of the H_2S and some of the CO_2. Amine treating is not effective for removal of mercaptan. In addition, amine treating cannot remove sufficient H_2S if the stream has to meet the copper strip corrosion test. For these reasons, caustic treating is the final polishing step employed downstream of amine units. Table 1-2 illustrates the chemistry of some of the important caustic reactions.

1.6.1 Sour Gas Absorber

An amine absorber (Figure 1-13) removes the bulk of H_2S from the sour gas. The sour gas leaving the sponge oil absorber usually flows into a separator which removes and liquefies hydrocarbon from vapors. The gas from the separator flows to the bottom of the H_2S contactor where it is met with a countercurrent flow of the cooled-lean amine

Table 1-2
**Acid/Base Reactions Encountered Most Frequently
by Oil Industry Caustic Treaters**

Carbon Dioxide		
CO_2 + 2 NaOH	$\rightarrow$	$Na_2CO_3 + H_2O$
Hydrogen Sulfide		
H_2S + 2 NaOH	$\rightarrow$	$Na_2S + 2\ H_2O$
Mercaptan Sulfur		
RSH + NaOH	$\rightarrow$	$RSNa + H_2O$
Naphthenic Acid		
RCOOH + NaOH	$\rightarrow$	$RCOONa + H_2O$

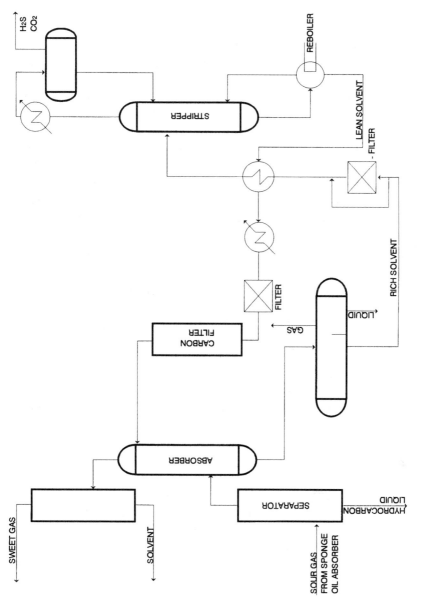

Figure 1-13. A typical amine treating system.

from an amine regenerator. The treated fuel gas leaves the top of the H$_2$S absorber and then goes to a settler drum for the removal of entrained solvent. The fuel gas then flows out of the settler drum to the battery limits.

The rich amine leaves the bottom of the H$_2$S contactor and goes to a flash separator for separation of dissolved hydrocarbons from the amine solution. The rich amine is pumped from the separator to the amine regenerator. A portion of the rich amine flows through a particle filter and a carbon bed filter. The particle filters are used to remove dirt, rust, iron sulfide, and other such material. The carbon filter, located downstream of the particle filters, removes residual hydrocarbons from the amine solution.

In the amine regenerator, the rich amine solution is heated to reverse the acid-base reaction that takes place in the contactor. The heat is supplied by a steam reboiler. The hot-lean amine is pumped from the bottom of the regenerator and exchanges heat with the rich amine in the lean-rich exchanger. From the lean-rich exchanger, the lean amine flows to a cooler before entering the contactor.

The sour gas containing small amounts of amine leaves the top of the regenerator. The vapor flows through a condenser and to the accumulator. The sour gas is sent to the sulfur unit while the condensed liquid is returned to the regenerator.

For many years, nearly all the amine units were using monoethanolamine (MEA) or diethanolamine (DEA). However, in recent years the use of tertiary amines such as methyl diethanolamine (MDEA) has been popular. These solvents are generally less corrosive and require less energy to regenerate.

1.6.2 LPG Treating

The LPG stream containing a mixture of C$_3$'s and C$_4$'s must be treated to remove hydrogen sulfide and mercaptan to produce a noncorrosive, less odorous, and less hazardous product. The C$_3$'s and C$_4$'s from the debutanizer accumulator flows to the bottom of the H$_2$S contactor. The operation of this liquid/liquid contactor is similar to that of the fuel gas absorber. In the LPG contactor, the amine is normally the continuous phase. In this mode, the amine-hydrocarbon interface is at the top of the contactor. This interface level controls the LPG flow out of the contactor. A number of liquid/liquid contactors are operated with the hydrocarbon as the continuous phase. In this case, the interface is controlled at the bottom of the contactor. The treated C_3's, C_4's stream leaves the top of the contactor where it is often installed to recover the carry-over amine from the treated LPG.

1.6.3 Caustic Treating

As stated earlier, amine treating is not effective for the removal of mercaptan. Mercaptans are organic sulfur compounds having the general formula of R-S-H. There are two options for treating mercaptans. One is sweetening them, i.e., converting them to disulfides and extracting the disulfide stream. The other option is to leave the converted disulfides in the product.

Sweetening of the FCC gasoline is usually sufficient to meet its sulfur specification. However, in areas where "reformulated" gasoline is marketed, regulations will call for a substantial reduction of total sulfur in the gasoline. The mercaptans in the LPG need to be extracted as required by the downstream processes such as alkylation. The presence of sulfur increases acid consumption and also produces undesirable by-products.

Both sweetening and extraction processes (Figure 1-14) commonly use caustic. The LPG and the gasoline may contain low ppm levels of H_2S. In this case, it is common practice to pass these streams through a caustic pre-wash drum to neutralize the H_2S.

The sweetening process utilizes a low-baumé caustic solution, an oxidizing catalyst, and air to convert mercaptans to disulfides in a mixing vessel. The reactions take place according to the following equations:

$$RSH + NaOH + catalyst \rightarrow RSNa + H_2O \qquad (1\text{-}10)$$

$$2RSNa + 1/2\ O_2 + H_2O + catalyst \rightarrow RSSR + 2NaOH \qquad (1\text{-}11)$$

The mixture of caustic and disulfides is transferred to a settler. From the settler, the treated gasoline flows to a coalescer, sand filter, or wash water tower before going to storage. The caustic solution is recirculated to the mixing vessel. In the extraction process, the LPG from the prewash tower enters the bottom of an extractor column. The

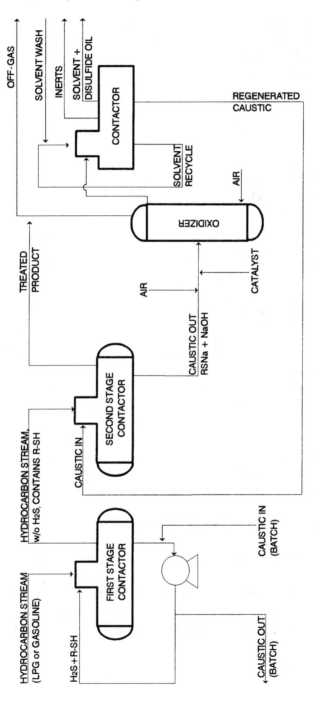

Figure 1-14. Caustic sweetening and extraction process. *(Adapted from Merichem Company, Houston, Texas.)*

extractor is a liquid/liquid contactor in which the LPG is counter-currently contacted by a caustic solution. The mercaptans are removed by way of Equation 1-10. The treated LPG leaves the top of the extractor and goes on to a settler, where entrained caustic is separated.

From the bottom of the extractor, the caustic solution containing sodium mercaptide enters the regenerator (oxidizer). Plant air and oxidizing catalyst are injected into the incoming caustic solution. The oxygen catalytically reacts with the sodium mercaptide to form disulfide oil as shown in Equation 1-11. The oxidizer overhead stream flows to a disulfide contactor/separator. A hydrocarbon solvent such as naphtha is circulated to physically wash the disulfide oil out of the regenerated caustic. In the disulfide separator, the gaseous by-products, namely nitrogen and excess oxygen, are vented. The regenerated caustic is returned to the extractor and the solvent containing disulfide oil is disposed in other units.

SUMMARY

Fluid catalytic cracking is one of the most important conversion processes in a petroleum refinery. The process incorporates most phases of chemical engineering fundamentals, such as fluidization, heat/mass transfer, and distillation. The heart of the process is the reactor-regenerator, where most of the innovations have occurred since 1942.

The objective of the FCC unit is to convert low-value, high-boiling-point feedstocks into more valuable products such as gasoline and diesel. FCC is an extremely efficient process. The cracking reactions result in deposition of coke on the catalyst. The catalyst activity is restored by burning off the coke with air. The catalyst-burning step supplies the heat for the reactions through circulation of catalyst between reactor and regenerator.

The hot-product vapors from the reactor are recovered in the main fractionator and the gas plant. The primary function of the main fractionator is to separate the heaviest products, such as light-cycle and decanted oil, from the gasoline and lighter products. The gas plant separates the main fractionator overhead vapors into gasoline, C_3's, C_4's, and fuel gas. These products contain sulfur compounds and need to be treated prior to being used. A combination of amine and caustic solutions are employed to sweeten these products.

REFERENCES

1. Williamson, M., "Worldwide Refining," *Oil & Gas Journal,* December 19, 1994, p. 55.
2. Strong, R. C., Majestic, V. K., and Wilhelm, S. M., "Basic Steps Lead to Successful FCC Corrosion Control," *Oil & Gas Journal,* September 31, 1991, pp. 81–83.

CHAPTER 2

FCC Feed Characterization

Refiners are faced with processing many different types of crude oil. As market conditions and crude quality fluctuate, so do cat cracking feedstock properties. Feed characterization is the process of determining physical and chemical properties of the feed. Two feeds with similar boiling point ranges may exhibit dramatic differences in cracking performances and product yields. Because of these variable conditions, often the only constant in FCC operations is the continual change in the feedstock quality.

FCC feed characterization is one of the most important requirements of cat cracking. Understanding feed properties and knowing their impact on unit performance are essential in anything that has to do with FCC operations, including troubleshooting, catalyst selection, unit optimization, and subsequent process evaluation. Feed characterization is a means of relating feed quality to product yields and qualities. Knowing the effects of a feedstock on unit yields, a refiner can purchase the feedstock that would maximize profitability. It is not uncommon for many refiners to purchase raw crude oils or FCC feedstocks without knowing their impact on unit operations. At times, this lack of knowledge can cause unit shutdowns for several weeks.

Sophisticated analytical techniques such as mass spectrometry are not practical for determining complete composition of FCC feedstocks on a routine basis. Empirical correlations are excellent alternatives for characterization of FCC feeds—they are easy to use and only require routine tests that commonly are performed by the refinery laboratory. These correlations do have their limitations: They are usually for an olefin-free feed, and they cannot distinguish among different paraffinic molecules or segregate an aromatic compound that may also contain a paraffinic and naphthenic structure group. Nevertheless, these correlations are very practical tools for tracking unit performance and for troubleshooting. Additionally, they are important in process design and catalyst research.

The two primary factors that affect feed quality are:

- Hydrocarbon Classification
- Impurities

2.1 HYDROCARBON CLASSIFICATION

The hydrocarbon types in the FCC feed are broadly classified as paraffins, olefins, naphthenes, and aromatics (PONA).

2.1.1 Paraffins

Paraffins are straight or branched chain hydrocarbons having the chemical formula C_nH_{2n+2}. The name of each member ends with *-ane;* examples are propane, isopentane, and normal heptane (Figure 2-1).

In general, FCC feeds are predominantly paraffinic. The paraffin content is typically between 50 wt% and 65 wt% of the total feed. Paraffinic stocks are easy to crack and normally yield the greatest amounts of total liquid products, the most gasoline, the lowest fuel gas, and the least octane number.

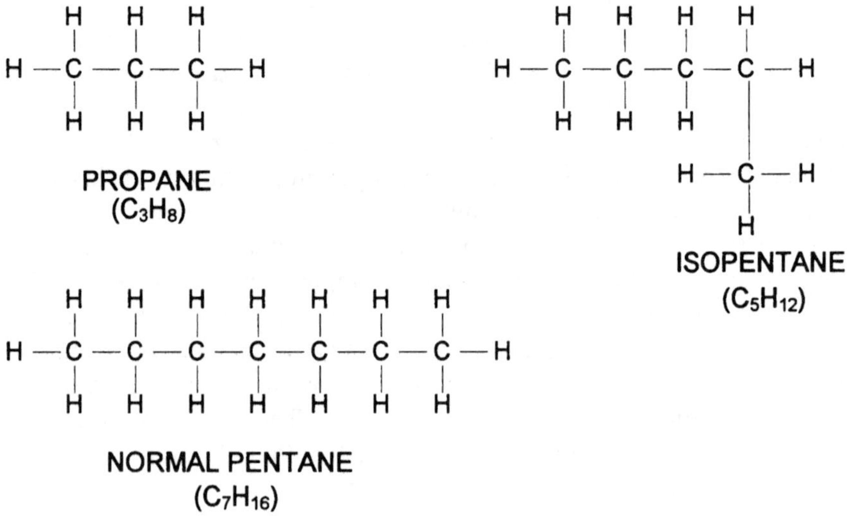

Figure 2-1. Examples of paraffins.

2.1.2 Olefins

Olefins are unsaturated compounds with a formula of C_nH_{2n}. The name of these compounds ends with -*ene*, such as ethane (ethylene) and propene (propylene). Figure 2-2 shows typical examples of olefins. Compared to paraffins, olefins are unstable and can react with themselves or with other compounds such as oxygen and bromine solution. Olefins do not occur naturally; they show up in the FCC feed as a result of preprocessing the feeds elsewhere. These processes include thermal cracking and other catalytic cracking operations.

Olefins are not the preferred feedstocks to an FCC unit. They usually crack to form undesirable products, such as slurry and coke. Typical olefin content of FCC feed is less than 5 wt% unless charging unhydrotreated thermally-produced gas oils.

2.1.3 Naphthenes

Naphthenes (C_nH_{2n}) have the same formula as olefins, but their characteristics are significantly different. Unlike olefins that are straight-chain compounds, naphthenes are paraffins that have been "bent" into a ring

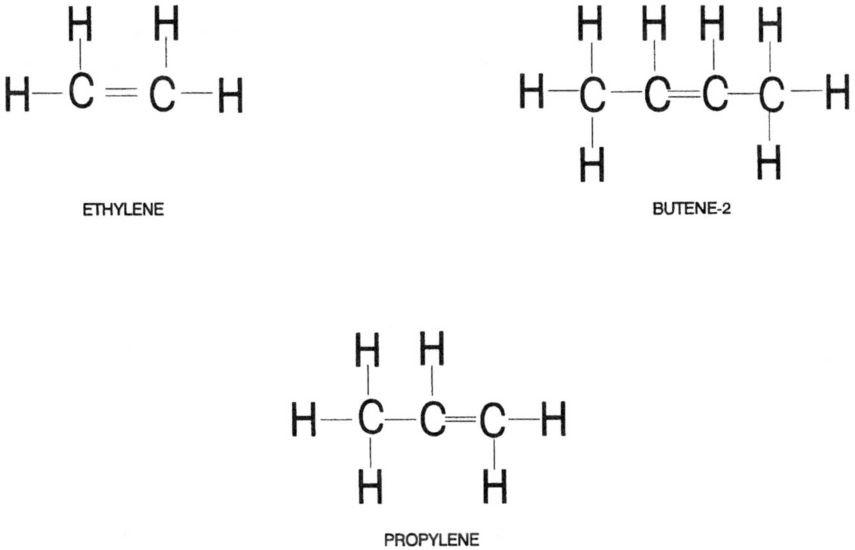

Figure 2-2. Examples of olefins.

or a cyclic shape. Naphthenes, like paraffins, are saturated compounds. Examples of naphthenes are cyclopentane, cyclohexane, and methyl-cyclohexane (Figure 2-3).

Naphthenes are desirable FCC feedstocks because they produce high-octane gasoline. The gasoline derived from the cracking of naphthenes has more aromatics and is heavier than the gasoline produced from the cracking of paraffins.

2.1.4 Aromatics

Aromatics (C_nH_{2n-6}) are similar to naphthenes, but they contain a stabilized unsaturated ring core. Aromatics (Figure 2-4) are compounds that contain at least one benzene ring. The benzene ring is very stable and does not crack to smaller components. Aromatics are not preferable as FCC feedstocks because most of the molecules will not crack. The cracking of aromatics mainly involves breaking off the side chains, and this can result in excess fuel gas yield. In addition, some of the aromatic compounds contain several rings (polynuclear aromatics) that can "compact" to form what is commonly called "chicken wire." Figure 2-5 illustrates an example of a polynuclear aromatic compound. Some of these compacted aromatics will end up on the catalyst as carbon residue (coke), and some will become slurry product. In comparison to paraffins, the cracking of aromatic stocks results in lower conversion, lower gasoline yield, and less liquid volume gain with higher gasoline octane.

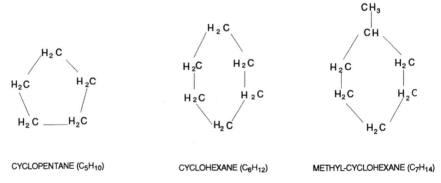

CYCLOPENTANE (C_5H_{10})   CYCLOHEXANE (C_6H_{12})   METHYL-CYCLOHEXANE (C_7H_{14})

Figure 2-3. Examples of naphthenes.

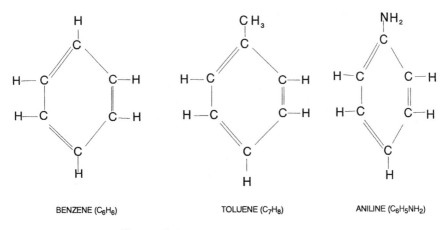

BENZENE (C_6H_6)   TOLUENE (C_7H_8)   ANILINE ($C_6H_5NH_2$)

Figure 2-4.   Examples of aromatics.

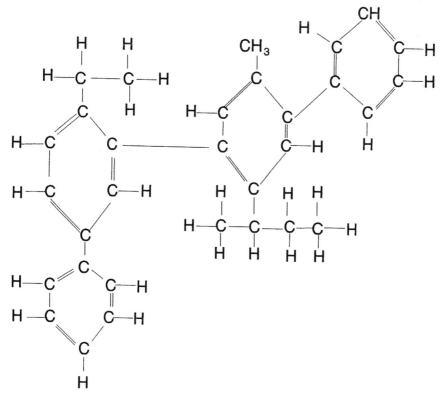

Figure 2-5.   An example of a polynuclear aromatic molecule.

2.2 FEEDSTOCK PHYSICAL PROPERTIES

To properly characterize an FCC feedstock, one must determine both its chemical and physical properties. Because sophisticated analytical techniques, such as mass spectrometry, are not practical in determining the chemical composition of an FCC feedstock, physical properties are often used to provide qualitative measurement of the feed's composition. The refinery laboratory is usually equipped to carry out theses physical property tests on a routine basis. The most widely used properties are:

- °API Gravity
- Distillation
- Aniline Point
- Refractive Index
- Bromine Number and Bromine Index
- Viscosity
- Conradson, Ramsbottom, and Heptane Insoluble

2.2.1 °API Gravity

The °API gravity measures the lightness or heaviness of a hydro-carbon liquid. The liquid specific gravity (SG) is another common term used in the conversion of volume to weight. The liquid SG is the relative weight of a volume of sample to the weight of the same volume of water at 60°F (15.5°C). The SG relates to °API gravity by the equations:

$$SG(@\,60°\,F) = \frac{141.5}{131.5 +° API_{Gravity}}$$

$$° API_{Gravity} = \frac{141.5}{SG(at\,60°\,F)} - 131.5$$

These equations show that the higher the °API gravity, the lighter the liquid sample. In petroleum refining, the measurement of °API gravity is widely practiced for virtually any feed or product stream. The ASTM D-287 is a very simple test typically performed by a lab technician or unit operator. The method involves inserting a glass hydrometer into a cylinder containing the sample and reading the °API gravity along with the fluid temperature indicated on the hydrometer scale. The °API gravity is always reported at 60°F (15.5°C). Standard tables similar to Table 2-1 can be used to convert the °API at any temperature back to 60°F.

For an FCC feed that is highly paraffinic (waxy), the sample should be heated to about 120°F (49°C) before immersing the hydrometer for testing. The heating step will ensure that the wax is melted, thereby eliminating erroneous readings.

The advantage of using °API gravity is that it magnifies small changes in the feed density. For example, going from a 24 °API to a 26 °API only changes the difference in specific gravity by 0.011 and the density by 0.72 lb/ft^3. However, a two-number shift in the feed °API gravity could have profound effects on the cat cracker's yields.

Daily monitoring of °API gravity provides the operator with a tool to predict changes in unit operation. For the same distillation range, the 26 °API feed is easier to crack than the 24 °API feed because the 26 °API feed has more long-chain paraffinic molecules. In contact with

Table 2-1
Example of Temperature Correction from Observed °API Gravity to °API Gravity @ 60°F

Temp °F	°API Gravity at Observed Temperature									
	26.0	26.1	26.2	26.3	26.4	26.5	26.6	26.7	26.8	26.9
	Corresponding °API Gravity at 60°F									
100	23.4	23.5	23.6	23.7	23.8	23.9	24.0	24.2	24.2	24.3
105	23.1	23.2	23.3	23.4	23.5	23.6	23.7	23.8	23.9	24.0
110	22.8	22.9	23.0	23.1	23.2	23.3	23.4	23.5	23.5	23.7
115	22.5	22.6	22.7	22.8	22.9	22.9	23.0	23.1	23.2	23.3
120	22.2	22.3	22.3	22.4	22.5	22.6	22.7	22.8	22.9	23.0
125	21.8	21.9	22.0	22.1	22.2	22.3	22.4	22.5	22.6	22.7
130	21.5	21.6	21.7	21.8	21.9	22.0	22.1	22.2	22.2	22.4
135	21.2	21.3	21.2	21.5	21.6	21.7	21.8	21.9	22.0	22.1
140	20.9	21.0	21.1	21.2	21.3	21.4	21.5	21.6	21.7	21.8

the 1300°F (704°C) catalyst, these molecules are easier to rupture into valuable products.

Long straight-chain paraffins are important to the economics of an FCC unit because as these molecules are cracked catalytically, the unit conversion is increased. Increases in gasoline and LPG rates are compensated by decreases in less desirable products such as slurry and fuel gas.

Results from the simple °API gravity test provide valuable information about the quality of a feed. The shift in feed °API usually signals changes in other feed properties such as carbon residue and aniline point. Therefore, additional tests may be needed to fully analyze the feed.

2.2.2 Distillation

Boiling point distillation data also provide information about the quality and composition of a feed, as shown later in this chapter. Before proceeding to analyze the data, it is important to understand different testing methods and to be aware of their limitations.

The feed to the cat cracker in a typical refinery is a blend of *gas oils* from such operating units as crude, vacuum, deasphalted, and coker. Some refiners purchase outside FCC feedstocks to keep the FCC feed rate maximized. Other refiners process varying amounts of residue in their cat crackers. In recent years, the trend has been toward cutting heavier in distillation and adding residue in the form of either atmospheric or vacuum bottoms. Residue is most commonly defined as the fraction of feed that boils above 1050°F (565°C).

Each FCC feed stream has different boiling point characteristics and compositions. The frequency and method of testing feed streams vary from one refiner to another. Some analyze daily, others two or three times a week, and some once a week. The frequency depends on how the distillation results are applied, the variation in crude slates, and the availability of lab personnel to conduct the tests.

The fractional distillation test conducted in the laboratory involves measuring the temperature of the distilled vapor at the initial boiling point (IBP), as volume percent fractions 5, 10, 20, 30, 40, 50, 60, 70, 80, 90, and 95 are collected, and at end point (EP). Three ASTM methods are currently used to measure boiling points: D-86, D-1160, and D-2887.

D-86 is the most common method used in refineries. The oil sample is distilled at atmospheric pressure. Samples with EP of less than 750°F (400°C) are normally tested with the D-86 method; above this temperature, the sample begins to crack. Thermal cracking is identified by a drop in the temperature of distilled vapor, the presence of brown smoke, and a rise in the system pressure. Additionally, above 750°F liquid temperature, the glass distilling flask begins to deform. Virtually all of today's FCC feeds are too heavy to use the D-86 method, therefore, the boiling points are obtained using either ASTM D-1160 or ASTM D-2887.

D-1160 is run under vacuum (one millimeter of mercury). The results are converted to atmospheric pressure manually using standard correlations. Some newer apparatuses have a built-in software program that does the conversion automatically. The application of D-1160 is limited to a maximum EP temperature of about 1000°F (538°C) at atmospheric pressure. Again, above this temperature, the sample begins to crack thermally.

D-2887 is a low-temperature *simulative distillation* (SIMDIS) method that measures the vol% of true boiling point cuts using gas chromatography (GC) principles. Like D-1160, its use is limited to a maximum end point temperature of about 1000°F (538°C). However, there are new GC-based systems on the market that can measure boiling point temperatures as high as 1350°F (750°C), and this development will be useful in determining boiling points of feedstocks that contain residue and in the characterization of raw crudes. Compared with D-1160, SIMDIS is less labor intensive, more reproducible, and generally more accurate at IBP and at the 5% and 10% points.

Distillation data provide information about the fractions of feed that boil at less than 650°F (343°C) and also the fractions that boil over 900°F (482°C). The inclusion of a light virgin feed, the fraction that boils under 650°F, often results in a greater LCO yield and thus a lower unit conversion than the corresponding heavier fraction. Sources of these fractions are atmospheric gas oil, vacuum light gas oil, and coker light gas oil. The reasons for observing a lower conversion of light virgin feed are as follows:

1. Lower molecular weight means less cracking.
2. Light processed stocks are very aromatic.
3. Light aromatics have fewer crackable side chains.

Economics and unit configuration of each refiner dictate whether or not to include the less than 650°F materials in the FCC feed. As a general rule, every effort should be made to minimize this fraction. Minor improvements in the operation of the distillation columns can substantially reduce the amount of light gas oil in the FCC feed. However, including light gas oil in FCC feed reduces the amount of coke laid on the catalyst. Less coke means a lower regenerator temperature. Consequently, light gas oil can be used as a "quench" to decrease the regenerator temperature and to increase catalyst-to-oil ratio.

The fractions that boil above 900°F (482°C) provide an indication of the coke-making tendency of a given feed. Associated with this 900°F+ fraction is a higher level of contaminants such as metals and nitrogen. As discussed in Section 2.3, these contaminants deactivate the catalyst and cause production of less liquid products and more coke and gas.

Distillation data is the backbone of FCC feed analyses. As shown in Section 2.4, published correlations use distillation data to determine chemical composition of the FCC feed.

2.2.3 Aniline Point

Aniline is an aromatic amine ($C_6H_5NH_2$) and, when used as a solvent, is selective in aromatic molecules. Aniline is used to determine aromaticity of FCC feedstocks. Aromatics are more soluble in aniline than paraffins and naphthenes. Aniline point (AP) is the minimum temperature for a complete miscibility of equal volumes of aniline and the sample.

The aniline point increases with paraffinicity and decreases with aromaticity. It also increases with molecular weight. Naphthenes and olefins show values that lie between those for paraffins and aromatics. Typically, an aniline point of greater than 200°F (93°C) indicates paraffinicity, and an aniline point of less than 150°F (65°C) implies aromaticity.

Aniline point is used mainly to estimate the aromaticity of gas oil and light stocks. Some correlations, such as TOTAL [1], use both aniline point and refractive index, whereas others, such as n-d-M [2], employ refractive index to characterize FCC feed.

Test method ASTM D-611 involves heating a 50/50 mixture of the feed sample and aniline until the two phases become well mixed. The test senses miscibility via a light source that penetrates through the sample. The mixture is then cooled, and the temperature at which the mixture becomes suddenly cloudy is the aniline point.

2.2.4 Refractive Index

Similar to aniline point, refractive index (RI) also shows how refractive or aromatic a sample is. The higher the RI, the less crackable the sample. A given feed having an RI of 1.5105 is more difficult to crack than a feed with an RI of 1.4990. The RI can be either measured in a lab (ASTM D-1218) or predicted using correlations such as the one published by TOTAL Petroleum.

In the laboratory, RI is measured using a refractometer. The technician spreads a small amount of sample on the faces of both prisms in the refractometer. The light is then directed at the sample and the scale is read. The observed scale is then converted to a refractive index value by the use of tables supplied with the instrument and the corresponding sample temperature.

Both refractive index and aniline point tests are employed to qualitatively measure aromaticity of a stock. With dark and viscous samples, both methods have their limitations. For darker samples, the aniline point test is slightly more accurate because of its larger scale over the same range of aromatics. There is no industry-wide agreement as to which method is more accurate—some prefer refractive index and others prefer aniline point. For instance, the three published correlations that will be discussed in Section 2-4 all use refractive index at 68°F (20°C) for calculations of feed composition. The problem is that at 68°F, most FCC feeds are solid and their refractive indexes cannot be determined accurately. Both the TOTAL and API [3] correlations predict RI values using feed properties such as specific gravity, molecular weight, and average boiling point.

2.2.5 Bromine Number and Bromine Index

Both bromine number (ASTM D-1159) and bromine index (ASTM D-2710) are qualitative methods to measure the reactive sites of a sample. Bromine reacts not only with olefin bonds but also with other molecules in the sample that have basic nitrogen and some aromatic sulfur derivatives. Nevertheless, olefins are the most common reactive sites and the bromine number is used to indicate olefinicity of the feed.

Bromine number is the number of *grams* of bromine that will react with 100 grams of the sample. Typical bromine number results are: less than 5 for hydrotreated feeds, 10 for vacuum heavy gas oil, and 50 for coker gas oil. A general rule of thumb is that the olefin fraction of the sample is equivalent to 1/2 of its bromine number.

Alternatively, bromine index is the number of *milligrams* of bromine that will react with 100 grams of the sample, and is used mostly by the chemical industry for stocks that have very low olefin contents.

2.2.6 Viscosity

Measuring viscosity provides information about chemical composition of a substance. As the viscosity of a sample increases, there is an increase in hydrogen content and a decrease in the fraction of aromatics.

Normal practice is to measure viscosity at two different temperatures: typically 100°F and 210°F. However, for many FCC feeds, the sample is too thick to flow at 100°F. In this case, the sample is heated to about 130°F. With two temperatures, a viscosity-temperature chart similar to the one shown in Appendix 1 can be used to extrapolate viscosity over a wide temperature range [4]; although viscosity is not a linear function of temperature, the coordinates on these charts have been adjusted to show the linear relationship between viscosity and temperature.

Viscosity of an oil is a measurement of its resistance to flow. Although the unit of absolute viscosity is *poise*, its measurement is difficult. Instead, kinematic viscosity is determined by measuring the pressure drop at a given flow through a capillary tube of specific diameter and length. The unit of kinematic viscosity is the *stoke*. However, in general practice, *centistoke* is used. Poise is related to stoke by the equation:

$$\text{Centistokes} = \frac{\text{Centipoise}}{\text{Density}}$$

The Saybolt viscosity (ASTM D-88) is the most popular method of measuring viscosity of an FCC feed. This method covers two procedures: one for the Saybolt Universal Viscometer (SUV) and the other for the Saybolt Furol Viscometer (SFV). Both procedures measure the time for a fixed volume of the sample to flow through a calibrated tube at a controlled temperature.

The characteristic difference between the two instruments is the inside diameter (ID) of the outlet tube. The SUV uses a 0.176-centimeter ID and the SFV uses a 0.315-centimeter ID. For samples that have flowing time of greater than 600 seconds, it is common practice to use the Furol Viscometer. For most conventional gas oils, the flowing time is short enough that the Universal Viscometer is frequently used. The tube dimensions are such that the Furol viscosity of oil is numerically one-tenth (1/10) of the Universal viscosity at the same temperature.

2.2.7 Conradson, Ramsbottom, and Heptane Insoluble

One area of cat cracking not fully understood is the proper determination of carbon residue of the feed and how it affects the unit's coke make. Carbon residue is defined as the carbonaceous resid formed after thermal destruction of a sample. Cat crackers are generally limited in coke burn-off capacity, therefore, the inclusion of residue produces more coke and forces a reduction in FCC throughput. The carbon residue of conventional gas oil feeds are generally less than 0.5 wt%, whereas in feeds containing residue the number could be as high as 15 wt%.

Three popular tests are presently used to measure carbon residue or concarbon of FCC feedstocks: Conradson, Ramsbottom, and Heptane Insoluble. Each test has some advantages and disadvantages, but none of them provide a rigorous definition of carbon residue or asphaltenes.

The Conradson test (ASTM D-189) measures carbon residue by evaporative and destructive distillation of the sample in a preweighed sample dish. The test involves applying heat to the sample using a gas burner until vapor ceases to burn and no blue smoke is observed. After cooling, the sample dish is reweighed and the percentage of carbon residue of the sample is calculated. The test, though popular, is not a good measurement of the coke-forming tendency of FCC feed because it is an indication of the thermal, rather than catalytic, property of the feed. In addition, the Conradson test is labor intensive and usually not reproducible, and the testing procedure tends to be subjective.

The Ramsbottom test (ASTM D-524) is also used to measure carbon residue. The test calls for introducing 4 grams of sample into a preweighed glass bulb and then inserting the bulb in a heated bath for

20 minutes. The bath temperature is maintained at 1027°F (553°C). After 20 minutes of heating, the sample bulb is cooled and reweighed. Compared with the Conradson test, Ramsbottom is more precise and reproducible; nevertheless, both methods produce similar results and often are interchangeable (see Figure 2-6).

The Heptane Insoluble (ASTM D-3279) method is commonly used to measure the asphaltene content of the feed. No one has a clear understanding of the asphaltene molecular structure. Asphaltenes are clusters of polynuclear aromatic sheets. The asphaltenes are insoluble in C_3 to C_7 paraffins. The amount of asphaltenes that precipitate varies from one solvent to another, so it is important that the reported asphaltene

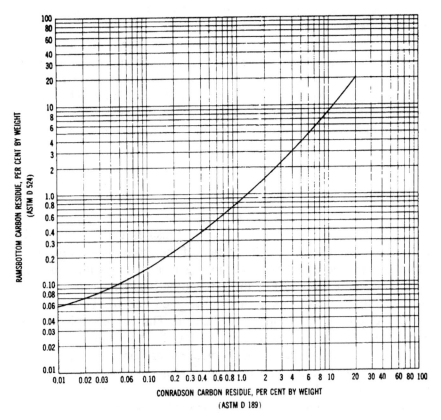

Figure 2-6. Ramsbottom Carbon Residue versus Conradson Carbon Residue. *(Copyright ASTM. Reprinted with permission.)*

values be identified with the appropriate solvent. Both normal heptane and pentane insolubles are widely used test methods for measuring asphaltenes. Although they do not provide rigorous definitions of asphaltenes, they provide practical ways of assessing coke formation of FCC feedstocks.

2.3 IMPURITIES

In recent years, refiners have been processing heavier crudes because the conversion of these heavy crudes to valuable products provides the refiners with financial benefits. The cat cracker, as the main conversion unit, is designed to handle a variety of feedstocks. Today's FCC feedstocks are generally heavier and contain higher levels of nitrogen, sulfur, and metals. These impurities have negative effects on unit performance. Understanding the nature and effects of these contaminants is essential in feed and catalyst selection as well as troubleshooting of the unit.

Most of the impurities in the FCC feed exist as components of large organic molecules. The most common contaminants are:

- Nitrogen
- Sulfur
- Nickel
- Vanadium
- Sodium

Except for sulfur, all these contaminants poison the FCC catalyst, causing it to lose its ability to produce valuable products. Sulfur in the feed increases operating costs because additional product treatment facilities are required to reduce the sulfur content to meet product specifications and comply with environmental regulations.

2.3.1 Nitrogen

Nitrogen in the FCC feed refers to *organic* nitrogen compounds. The nitrogen content of FCC feed is often reported as *basic* and *total* nitrogen. The total nitrogen is the sum of basic and *nonbasic* fractions. Basic nitrogen is about one fourth to one half of total nitrogen.

The word "basic" denotes having the capability of reacting with acids. FCC catalysts have acid sites that these basic nitrogen compounds will neutralize. The result is a temporary loss of catalyst activity and a subsequent drop in unit conversion (Figure 2-7). In the regenerator, the nitrogen is converted predominantly to nitrogen oxide (NO_x). The NO_x leaves the unit with the flue gas. The burning of nitrogen in the regenerator restores the activity of the catalyst.

It has been my experience that there is little or no loss in unit conversion as long as the total and basic nitrogen content of the feed is less than 800 ppm and 300 ppm, respectively. This surely depends on the type of catalyst being used. An FCC catalyst with high zeolite content and an active matrix shows more tolerance to nitrogen content than a low-zeolite, low-active-matrix catalyst.

Directionally, lowering temperature of the feed to the riser will help to reduce the negative effects of nitrogen. At higher catalyst-to-oil ratio, more catalyst is available to absorb nitrogen, and therefore, nitrogen has lesser impact on the unit conversion.

For some refiners, hydrotreating the feed may be an appropriate economical approach. Except for most of the Californian and a few other crudes, FCC feeds with high nitrogen concentration also have other impurities. Therefore, it is difficult in a commercial unit to evaluate deleterious effects of nitrogen alone. Hydrotreating the feed not only decreases the nitrogen content but also can remove most other contaminants.

Aside from catalyst poisoning, nitrogen in the FCC feed is detrimental to the unit operation in several other areas. First, in the riser, some of the nitrogen is converted to ammonia and cyanide (H-CN). Cyanide accelerates the corrosion rate of the FCC gas plant equipment by removing the protective sulfide scale and exposing bare metals to further corrosion. This corrosion is the source of atomic hydrogen that ultimately results in hydrogen blistering. Cyanide formation tends to increase with cracking severity. Second, some of the nitrogen compounds will end up in FCC light cycle oil (LCO). These compounds, such as pyrolles and pyridines [5] are easily oxidized and will affect color stability. The amount of nitrogen that stays in the LCO depends on the conversion level. An increase in conversion results in a decreased percentage of nitrogen in the LCO and an increased percentage of nitrogen on the catalyst.

The source and gravity range of raw crude greatly influence the amount of nitrogen in the FCC feed (Table 2-2). Generally speaking, heavier crudes contain more nitrogen than the lighter crudes. In addition,

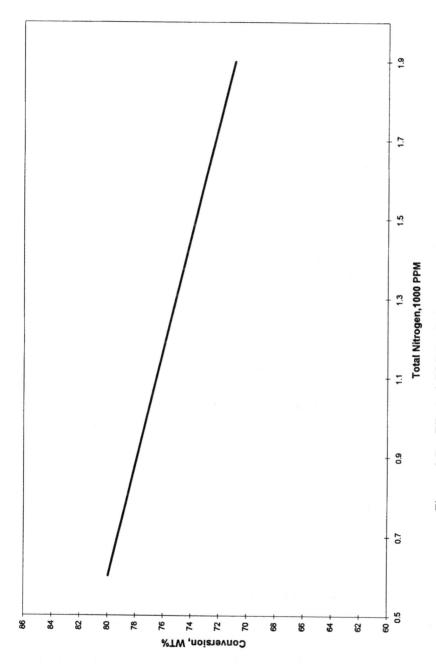

Figure 2-7.   Effect of FCC feed nitrogen on unit conversion.

Table 2-2
°API Gravity, Residue, and Nitrogen Content of Typical Crudes

Crude Source	°API Gravity	Vacuum Bottoms, vol%	Total Nitrogen* of Heavy Vacuum Gas Oil, PPM
Maya	21.6	33.5	2498
Alaska North Slope (ANS)	28.4	20.4	1845
Arabian Medium	28.7	23.4	829
Forcados	29.5	7.6	1746
Cabinda	32.5	23.1	1504
Arabian Light	32.7	17.2	1047
Bonny Light	35.1	5.3	1964
Brent	38.4	11.4	1450
West Texas Intermediate Cushing (WTIC)	38.7	10.6	951
Forties	39.0	10.1	1407

Nitrogen level varies with crude source and residue content.

nitrogen tends to concentrate in the residue portion of the crude. Figure 2-8 shows examples of nitrogen compounds found in crude oil.

The UOP test method 313 is commonly employed to determine the *basic* nitrogen content of FCC feed. This method calls for titrating the feed sample with perchloric acid using a 50/50 mixture of acetic acid and the sample. ASTM Method D-3228 (or chemiluminescent nitrogen detector) is used to measure the *total* nitrogen and involves converting all the nitrogen in the feed to ammonia and then titrating it with standard sulfuric acid.

2.3.2 Sulfur

FCC feedstocks contain sulfur in the form of organic-sulfur compounds such as mercaptan, sulfide, and thiophenes. Total sulfur in FCC feed is determined by using the X-ray spectrographic method (ASTM D-2622). The results are expressed as elemental sulfur.

Although desulfurization is not a main goal of cat cracking operations, approximately 50% of sulfur in the feed is converted to H_2S. In addition, the remaining sulfur in the FCC products are lighter and

A. **Neutral N – Compounds**

N
H
Indole

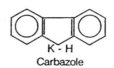

K - H
Carbazole

B. **Basic N – Compounds**

N
Pyridine

N
Quinoline

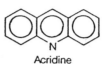

N
Acridine

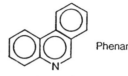

Phenanthridine.

N

C. **Weakly Basic N – Compounds**

N   OH
Hydroxiquinoline

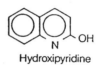

N   OH
Hydroxipyridine

Deriviates with R = H, alkyl-, phenyl-, naphthyl-

Nitrogen Distribution in Several Middle Eastern Oils

Content: 20-25% of nitrogen in 225-540°C gas oil fraction.
75-80% of nitrogen in 540°C plus vacuum resid fraction.

Type: 225-540°C gas oil fraction: 50% of nitrogen as neutral nitrogen compounds; 33% as basic, 17% as weakly basic.

540°C plus vacuum resid fraction: 20% of nitrogen in asphaltenes, 33% as neutral, 20% as basic, 27% as weakly basic.

Figure 2-8. Types of nitrogen compounds in crude oil [12].

can be desulfurized by low-pressure hydrodesulfurization processing. The H_2S is formed principally by the catalytic decomposition of non-thiophenic sulfur compounds. Table 2-3 shows the effects of feedstock sulfur compounds on H_2S production.

Table 2-3
Effects of Feedstock Sulfur Compounds on H$_2$S Production

Cracking Conditions: 7 Cat/Oil Ratio, 950°F, Zeolite Catalyst			
Feed Source	Conversion Vol%	% of Feed Sulfur Which Is Mercapcan or Sulfide and Not Aromatic in Nature	Vol% of Sulfur Converted* to H$_2$S
Mid Continent	72	38	47
West Texas	69	33	41
Coker Gas Oil	56	30	35
Hydrotreated West Texas	77	12	26
Heavy Cycle Oil	50	6	16

The % sulfur converted to H$_2$S depends largely on the type of sulfur in the feed and the residence time of the hydrocarbons in the riser.
Source: Wollaston [6]

As with H$_2$S, the distribution of sulfur among the other FCC products also depends on several factors, which include feed, catalyst type, conversion, and other operating conditions. Feed type and residence time are the most significant variables. Sulfur distribution in FCC products of several feedstocks is shown in Table 2-4. Figure 2-9 also illustrates the sulfur distribution as a function of the unit conversion. Frequently, as the residue content of crude oil increases, so does its sulfur content (Table 2-5).

For nonhydrotreated feeds and constant conversion (78 vol%), about 50 wt% of the sulfur in the feed is converted to hydrogen sulfide (H$_2$S). The remaining 50% of the sulfur is distributed approximately as follows: 6 wt% in gasoline, 23 wt% in light cycle oil, 15 wt% in decanted oil, and 6 wt% in coke. As shown in Table 2-6, the addition of residue to the feed increases the sulfur content of coke proportional to the incremental sulfur in the feed. Thiophenic sulfur compounds crack more slowly, and the uncracked thiophenes end up in gasoline, light cycle oil, and decanted oil.

In the hydrotreated feeds, more of the feed sulfur goes to coke, decanted oil, and light cycle oil than in the nonhydrotreated feeds. Hydrotreating reduces the sulfur content of all the products. The same sulfur atoms that were converted to H$_2$S in the FCC process are also

(text continued on page 59)

Table 2-4
Sulfur Distribution in FCC Products

	Feedstock Sources			
Feedstock	W. Texas Virgin Gas Oil	W. Texas Virgin Gas Oil (HDT)	Calif. Gas Oil	Kuwait DAO & Gas Oil Blend (HDT)
Sulfur Content, wt%	1.75	0.21	1.15	3.14
Conversion, vol%	77.8	77.8	78.7	80.1
Sulfur Distribution, Wt% of Feed Sulfur				
H_2S	42.9	19.2	60.2	50.0
Light Gasoline	0.2	0.9	1.6	1.9
Heavy Gasoline	3.3	1.9	7.9	5.0
LCO	28.0	34.6	20.7	17.3
Decanted Oil	20.5	34.7	6.8	15.3
Coke	5.1	8.7	2.8	10.3

Source: Huling [7]

Table 2-5
°API Gravity, Residue, and Sulfur Content of Some Typical Crudes

Crude Source	°API Gravity	Vacuum Bottoms, vol%	Sulfur Content of Vacuum Gas Oil, Wt%*
Maya	21.6	33.5	3.35
Alaska North Slope (ANS)	28.4	20.4	1.45
Arabian Medium	28.7	23.4	3.19
Forcados	29.5	7.6	0.30
Cabinda	32.5	23.1	0.16
Arabian Light	32.7	17.2	2.75
Bonny Light	35.1	5.3	0.25
Brent	38.4	11.4	0.63
West Texas Intermediate Cushing (WTIC)	38.7	10.6	0.63
Forties	39.0	10.1	0.61

Sulfur level varies with crude source and residue content.

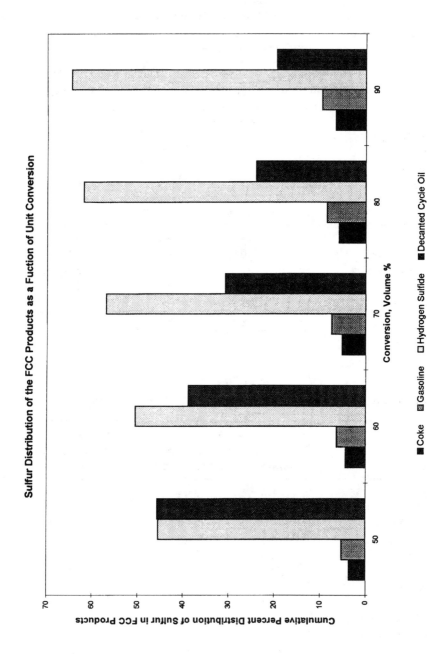

Figure 2-9. Sulfur distribution of the FCC products as a function of unit conversion.

(text continued from page 56)

being removed first in the hydrotreating process. The remaining sulfur compounds are harder to remove. The heavier and more aromatic the feedstock, the greater the level of sulfur in the coke (Table 2-7).

It should be emphasized that although hydrotreating increases percent sulfur in the coke and slurry, the actual amount of sulfur is substantially less than in the nontreated feeds. All other things being equal, sulfur plays a minor role in unit conversion and yields. Some sulfur

Table 2-6
Sulfur Content of Coke vs. Quantity of Residue in FCC Feed*

Pilot Plant Data, Riser Cracking for Maximum Liquid Recovery		
Feedstock Type	Feed Sulfur, Wt%	Sulfur in Coke Wt% of Feed
Gas Oil	0.7	3.5
Gas Oil + 10% of West Texas Sour VTB	1.0	13.8
Gas Oil + 10% of West Texas Sour VTB	1.32	18.6

As the residue content of the feed is increased, there is a marked increase in the coke's sulfur due to higher coke yield and a higher sulfur content of the coke precursors.
Source: Campagna [8]

Table 2-7
Sulfur Content of Coke vs. Hydrotreated* FCC Feed Quality

Pilot Plant Data, Riser Cracking for Maximum Liquid Recovery			
Feedstock Source	Feedstock Sulfur, wt%	Hydrocarbon Type % Tri-aromatics*	Sulfur in Coke, Wt% of Feed
Light Arabian HDS	0.21	7.3	28.1
Heavy Arabian HDS	0.37	17.6	48.2
Maya HDS	0.70	5.0	43.7

In a hydrotreated feed, the more polyaromatic type sulfur compounds, the more sulfur ends up in coke.
Source: Campagna [8]

in the form of aromatics does not convert to gasoline; it becomes predominantly cycle oil, depending on side-chain length, and also produces fuel gas. This tends to lower conversion and reduce maximum yields.

2.3.3 Metals

Metals, such as nickel, vanadium, and sodium, are present in crude oil. These metals are concentrated in the heavy boiling range of atmospheric or vacuum residue, unless they are carried over with the gas oil by entrainment.

These metals are catalysts themselves and therefore promote some undesirable reactions, such as dehydrogenation and condensation. Dehydrogenation means the removal of valuable hydrogen from the feed. Condensation means the formation of "chicken wire" aromatic molecules, as previously discussed. The results are increased hydrogen and coke yields and lower gasoline yields. These metals permanently poison the FCC catalyst by lowering the catalyst activity, thereby reducing its ability to produce the desired products. Virtually all the metals in the FCC feed are deposited on the cracking catalyst. Feeds that contain more nickel than vanadium tend to be more paraffinic. The negative effects of each metal are discussed below.

Nickel (Ni)

As discussed in Chapter 3, there are two main parts to an FCC catalyst: the nonframework structure called matrix and the crystalline structure called zeolite. When FCC feed contacts the catalyst, the nickel in the feed deposits on the matrix. The nickel promotes dehydrogenation reactions, which remove hydrogen from stable compounds and leave behind heavy hydrocarbon molecules. These reactions result in higher hydrogen and coke yields. The higher coke content will result in a higher regenerator temperature which lowers the catalyst-to-oil ratio and causes loss of conversion at constant preheat temperature.

High nickel levels are normally encountered when processing heavy feed. Neither excess hydrogen nor excess regenerator temperature is desirable. Excess hydrogen is undesirable because it limits the capacity of the wet gas compressor, forcing a reduction in unit charge or lowering conversion.

A number of indices have been developed to relate metal activity to hydrogen and coke production. These indices predate the use of metal passivation in the FCC process. The most commonly used index is 4 × Nickel + Vanadium. This indicates that nickel is four times as active as vanadium in producing hydrogen. Other indices [9] used are as follows:

Jersey Nickel Equivalent Index = $1000 \times (Ni + 0.2 \times V + 0.1 \times Fe)$

Shell Contamination Index = $1000 \times (14 \times Ni + 14 \times Cu + 4 \times V + Fe)$

Davison Index = $Ni + Cu + \dfrac{V}{4}$

Mobil = $Ni + \dfrac{V}{4}$

These indices convert all metals to a common basis, generally either vanadium or nickel.

Nickel and other metals are most active as soon as they deposit on the catalyst. With time, they lose their initial effectiveness through continuous oxidation-reduction cycles. On the average, about one third of the nickel on the equilibrium catalyst will have the activity to promote dehydrogenation reactions.

A small amount of nickel in the FCC feed has a significant negative influence on the unit operation. In a "clean" gas oil operation, the nickel content of the feed is less than 0.5 ppm. At this level, the net hydrogen yield is about 30 to 40 standard cubic feet (scf) per barrel of feed which is a manageable rate that just about every FCC unit can handle. If the nickel level increases to 1.5 ppm, in a 50,000 barrel/day unit, this corresponds to a mere 16 pounds per day of nickel. However, the hydrogen make increases to 80–100 scf per barrel of feed. Chances are that unless the catalyst addition rate is increased or the nickel in the feed is passivated (see Section 3.7), the unit feed rate or the conversion needs to be reduced. This is because the produced gas will become lean and thus limit the pumping capacity of the wet gas compressor.

In most units, the effects of hydrogen do not result in additional coke yield. As discussed in Chapter 5, the coke yield in a cat cracker is constant. The coke yield does not go up, because other unit constraints such as a regenerator temperature and/or wet gas compressor would force the unit operator to either cut back on charge or reduce severity. Another drawback of high hydrogen yield is that it affects the recovery of C_3+ components in the gas plant. This is due to a drop in liquid-vapor ratio in the absorbers.

On a wt% basis, the increase in hydrogen is negligible. It is the sharp increase in the volume of the gas that drastically impacts unit performance. The composition of cracking catalyst has a noticeable impact on hydrogen yields. Catalysts with an active alumina matrix tend to increase the dehydrogenation reactions. In addition, the presence of chlorides in the feed also reactivates the aged nickel and thus results in large hydrogen yield.

There are two common means of tracking effects of nickel on the catalyst. These are hydrogen/methane ratio and standard cubic feet of hydrogen per barrel of feed. Aside from dehydration reactions, H_2/CH_4 is also more sensitive to reactor temperature and the type of catalyst than hydrogen yield alone. Therefore, measuring standard cubic feet of hydrogen per barrel of fresh feed is a better indicator of nickel activity than the H_2/CH_4 ratio. The typical H_2/CH_4 ratio for a gas oil having less than 0.5 ppm nickel is between 0.25 to 0.35 mole ratio. The equivalent H_2 make is between 30 and 40 scf/bbl of feed.

Frequently, it is more accurate to back-calculate the feed metals from the equilibrium catalyst inspection data than to analyze the feed randomly. Depending on the unit's constraints and economic margins, it is beneficial to use some type of passivation if the nickel on the equilibrium catalyst is greater than 1500 ppm.

Iron is usually present in FCC feed as tramp iron and is not catalytically active. Tramp iron refers to various iron oxides formed as part of corrosion by-products. However, copper is as active as nickel but its magnitude in the feed is substantially lower than nickel.

Vanadium

Like nickel, vanadium also promotes dehydrogenation reactions. But vanadium contributions to hydrogen yield are anywhere from 25 to 50 percent of those of nickel, and problems due to vanadium are more severe. Unlike nickel, vanadium does not stay on the surface of the catalyst. Instead, vanadium migrates to the inner (zeolite) part of the catalyst and destroys the zeolite frame structure, causing loss of catalyst surface area and activity. FCC regenerator operating conditions influence the severity of vanadium deactivation, as shown in Figure 2-10. This zeolite destruction is enhanced by the presence of sodium in the catalyst and severe regenerator temperature.

Vanadium is in FCC feed as part of organo-metallic molecules of high molecular weight. When these heavy molecules are cracked in the riser, they leave some coke residue containing vanadium on the catalyst. During regeneration, the coke is burned off and the vanadium is converted to vanadium oxides such as vanadium pentoxide (V_2O_5). V_2O_5 is highly mobile and can go from one particle to another. It also has a melting point of 1247°F (675°C) which allows it to destroy zeolite under typical regenerator temperature conditions.

There are several theories about the chemistry of vanadium poisoning. The prominent one involves conversion of V_2O_5 to vanadic acid (H_3VO_4) under regenerator conditions. Vanadic acid destroys the zeolite by hydrolysis of zeolite framework. The presence of sodium helps in lowering the melting point of the eutectic and accelerates zeolite collapse.

Alkaline earth metals

Alkaline earth metals in general and sodium in particular are detrimental to the FCC catalyst. Sodium in the riser permanently deactivates

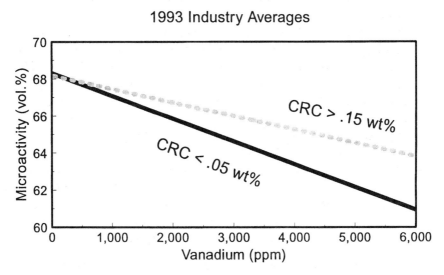

Figure 2-10. Vanadium deactivation varies with regenerator severity [13].

the catalyst by neutralizing its acid sites. In the regenerator, it causes the zeolite to collapse, particularly in the presence of vanadium.

There are two prime sources of sodium: sodium in the fresh catalyst and added sodium. Sodium exists in the fresh catalyst as part of the catalyst manufacturing process. Added sodium is the sodium that comes in with the fresh feed. For all practical purposes, the adverse effects of sodium are the same regardless of its origin. Nevertheless, this section discusses the effects of added sodium. Section 3.3 discusses the drawbacks of sodium that is inherent in the fresh catalyst.

Sodium in the FCC feed originates from the following places:

- *Caustic that is added downstream of the crude oil desalter.* Caustic injection downstream of the desalter is practiced to control crude unit overhead acid concentrations. Sodium chloride is thermally stable at the temperature found in the crude and vacuum unit heaters. This results in sodium chloride being present in either atmospheric or vacuum resids. Normally, refiners discontinue caustic injection while they process residue to the FCC unit.
- *Water soluble salts that are carried over due to inadequacy of the desalter operation.* An effective desalting operation is necessary when processing heavy feedstocks to the cat cracker. Chloride salts are usually water soluble and are removed from raw crude in the desalter. However, some of these salts can be carried over with the desalted crude.
- *Process of the refinery "slop."* A number of refiners process the refinery slop in their desalter. Depending on the amount and quality of slop, operation of the desalter is adversely affected. This results in potential carryover of salts with the desalted crude.
- *Purchased FCC feedstock that is brought to the refinery by tankers and barges that use salt water as ballast.* Outside FCC feedstocks are brought into the refinery by barges and tankers that use salt water as ballast.
- *The use of atomizing steam and/or water that contain sodium.* Just about every refiner practices some type of feed atomizing using either steam or water. The steam or water used in the atomization process can contain varying amounts of sodium depending on the quality of water treatment used in the refinery.

Another problem associated with sodium appears in the form of sodium chloride. Chlorides tend to reactivate aged metals by redistributing the metals on the equilibrium catalyst and allowing them to cause more damage.

The metals in the FCC feed have many deleterious effects. Nickel causes excess hydrogen production, forcing eventual loss in the conversion or thruput. Both vanadium and sodium destroy catalyst structure, causing loss in activity and selectivity. There are two general approaches to solving the undesirable effects of metal poisoning: increasing makeup rate of fresh catalyst or employing some type of metal passivation. Other schemes practiced by some refiners involve adding good quality equilibrium catalysts for flushing the metals and employing demetalization technology to remove metals from equilibrium catalyst. Both demetalization and passivation technology are addressed in Chapter 3.

2.4 EMPIRICAL CORRELATIONS

The typical refinery laboratory is not equipped to conduct PONA and other chemical analyses of the FCC feed on a routine basis. However, they are able to measure physical properties such as the °API gravity and distillation. As a result, empirical correlations have been developed by the industry to determine chemical properties from physical analyses.

The purpose of characterizing FCC feed is to provide quantitative and qualitative estimates of the FCC unit's performance. Process modeling of FCC employs the feed properties discussed thus far to predict FCC yields and product qualities. The benefit of the process model is its use in daily unit monitoring, catalyst evaluations, optimization, and process studies.

There are no standard correlations. Some companies have their own correlations that are considered proprietary, but this does not mean that these correlations do a better job at predicting yields. Nonetheless, they all incorporate most or some of the same physical properties. The most widely published correlations in use today are:

- K Factor
- TOTAL
- n-d-M Method
- API Method

2.4 K Factor

The K factor is a useful indication of feed crackability. It is normally calculated using feed distillation and gravity data, and measures aromaticity relative to paraffinicity. Higher K values indicate increased paraffinicity and more crackability. Any feed with a K value above 12.0 is considered paraffinic; a K value below 11.0 represents aromatic.

Like aniline point, the K factor differentiates between the highly paraffinic and aromatic stocks. However, within the narrow range (K = 11.5–12.0), the K factor does not correlate between aromatics and naphthenes. Instead, it relates fairly well to the paraffin content (Figure 2-11). Additionally, the K factor does not provide information as to how naphthene and paraffin contents relate to one another. As illustrated in Table 2-8, the ratio of naphthenes to paraffins can vary considerably, though K values might be the same for a number of feedstocks.

There are two widely used methods to calculate the K factor: the *Watson K* and the *UOP K*. The equations to calculate both Watson K and UOP K are as follows:

$$K(\text{Watson}) = \frac{(\text{MeABP} + 460)^{1/3}}{\text{SG}}$$

$$\text{UOP } K = \frac{(\text{CABP}) + 460)^{1/3}}{\text{SG}}$$

$$\text{UOP } K = \frac{(\text{VABP} + 460)^{1/3}}{\text{SG}}$$

$$\text{VABP} = \frac{(T(10\%) + T(30\%) + T(50\%) + T(70\%) + T(90\%))}{5}$$

$$\text{MeABP} = \frac{(\text{MABP}) + (\text{CABP})}{2}$$

$$\text{MABP} = \sum (f_{mi} \times TB_i)$$

$$\text{CABP} = \sum (f_{vi} \times TB_i^{1/3})^3$$

$$\text{SG} = \frac{141.5}{131.5 + \text{°API}}$$

Where:

MeABP = Mean Average Boiling Point, °F
SG   = Specific Gravity at 60°F
CABP = Cubic Average Boiling Point, °F

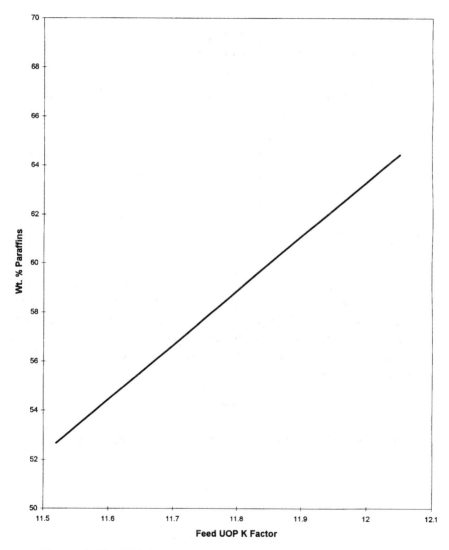

Figure 2-11. Weight percent paraffins at various UOP K factors.

VABP = Volumetric Average Boiling Point, °F
MABP = Molar Average Boiling Point, °F
f_{mi} = Mole Fraction of Component i
TB_i = Normal boiling point of pure component i, °F
f_{vi} = Volume fraction of component i
T = Temperature, °F

Table 2-8
Variation of C_N/C_P as a Function of K Factor*

Sample No.	UOP K Factor	$C_A + C_N$ (wt%)	C_N/C_P
1	11.70	46	0.47
2	11.69	45	0.44
3	11.70	46	0.44
4	11.67	45	0.43
5	11.70	45	0.39
6	11.70	44	0.35
7	11.70	42	0.33

The K factor relates well to aromatics + naphthenes, but not to naphthenes.
Source: Andreasson [10]

The UOP method uses CABP which, for all practical purposes, is the same as VABP as shown in Appendix 2. The UOP K factor is more popular than Watson K because the VABP data are readily available. The use of MeABP in the Watson method generally results in a lower K value than that of UOP. Example 2-1 illustrates steps to calculate the UOP K and Watson K factors.

Example 2-1

Determine UOP K and Watson K using the following FCC feed properties:

Feed Properties

°API Gravity	@ 60°F	23.5
SG	@ 60°F	0.913
Density	@ 20°C (68°F)	0.90
Refractive Index	@ 67°C	1.4810

Viscosity	@ 130°F, SUS	137.0
	@ 210°F, SUS	50.0
Sulfur, wt%		0.48
Aniline Point, °F(°C)		192.0 (88.9)

D-1160 @ 1 atm

Vol%	Temp. °F
10	652
30	751
50	835
70	935
90	1080

Procedure

1. Calculate VABP from distillation data.
2. Calculate the 10%–90% slope.
3. Calculate MeABP and CABP by adding corrections from Appendix 2 to VABP.
4. Determine Watson and UOP K.

Step 1: VABP = 1/5(652 + 751 + 835 + 935 + 1080)

VABP = 851°F = 455°C = 728.2°K

Step 2: 10%–90% slope

$$\text{Slope} = \frac{T(90\%) - T(10\%)}{80} = \frac{1080 - 652}{80}$$

Slope = 5.3 °F/percent off

Step 3: From Appendix 2, corrections to VABP are approximately –34°F for MeABP and –10°F for CABP.
Therefore:

MeABP = 851 – 34 = 817°F = 436°C

CABP = 851 – 10 = 841°F = 449.4°C

Step 4: Watson K $= \dfrac{(817 + 460)^{1/3}}{0.913} = 11.88$

$$\text{UOP K} = \frac{(841 + 460)^{1/3}}{0.913} = 11.98$$

Instead of using Appendix 2, the MeABP can be determined from the equation below [11]:

$$\text{MeABP} = \text{VABP} + 2 - \left(\frac{(T_{90} - T_{10})}{(170 + 0.075 \times \text{VABP})} + 1.5\right)^3$$

$$\text{MeABP} = 851 + 2 - \left(\frac{1080 - 852}{170 + 0.075 \times 851} + 1.5\right)^3$$

$$\text{MeABP} = 816°\text{F}$$

In the absence of full distillation data, the K factor can be estimated using the 50% point in place of MeABP.

In summary, the K factor is a simple tool that can provide information about the extreme aromaticity or paraffinicity of the feed. However, within the narrow range (K = 11.5 – 12.0), it cannot differentiate between ratio of paraffins, naphthenes, and aromatics. To determine these ratios, other correlations, such as TOTAL or n-d-M, should be employed.

2.4.2 TOTAL

The TOTAL correlations calculate aromatic carbon content, hydrogen content, molecular weight, and refractive index using routine laboratory measurements of feed properties. The correlations were developed from multiple linear regression of 33 feedstocks. Appendix 3 contains the formulas developed by TOTAL. Example 2-2 illustrates the use of TOTAL correlations.

Example 2-2

Using feed properties of Example 2-1, calculate MW, RI(20), C_A, and H_2 content of the feed.

Molecular Weight (MW)

$$MW = 0.0078312 \times (0.913)^{-0.0978} \times (88.9)^{0.1238} \times (455)^{1.6971}$$

$$MW = 446$$

Refractive Index (RI) @ 20°C

$$RI(20) = 1 + 0.8447 \times 0.913^{1.2058} \times 728.2^{-0.0557} \times 446^{-0.0044}$$

$$RI(20) = 1.5105$$

Refractive Index (RI) @ 60°C

$$RI(60) = 1 + 0.8156 \times (0.913)^{1.2392} \times (728.2)^{-0.0576} \times (446)^{-0.0007}$$

$$RI(60) = 1.4963$$

Hydrogen (H_2) Content, wt%

$$H_2 = 52.825 - 14.26 \times 1.5105 - 21.329 \times 0.913 - 0.0024$$
$$\times 446.5 - 0.052 \times 0.48 + 0.757 \times \ln (7.37)$$

$$H_2 = 12.23 \text{ wt\%}$$

Aromatic (C_A) Content, wt%

$$C_A = -814.136 + 635.192 \times 1.5105 - 129.266 \times 0.913 + 0.013$$
$$\times 446.5 - 0.34 \times 0.48 - 0.872 \times \ln(7.37)$$

$$C_A = 19.19 \text{ wt\%}$$

For FCC feeds, particularly the ones containing residue, the TOTAL correlation is more accurate at predicting aromatic carbon content than the n-d-M correlation. Table 2-9 illustrates this comparison. One option would be to calculate MW, RI(20), CA and H_2 from the TOTAL correlation, and use either n-d-M or API method to calculate the naphthene (C_N) and paraffin (C_P) content of the feed.

Table 2-9
Comparison of TOTAL Correlations with Other Methods

Correlation	Average Deviation	Absolute Average Deviation	Bias Maximum Deviation
Carbon Content			
n-d-M	5.14	4.67	12.99
API	2.88	2.53	9.13
TOTAL	0.93	0.00	3.45
Hydrogen Content			
Linden	0.31	−0.05	1.57
Fein-Wilson-Winn	0.36	0.19	1.43
Modified Winn	0.19	0.07	0.86
TOTAL	0.10	0.00	0.42
Molecular Weight			
API	62.0	−62.0	180.9
Maxwell	63.3	−63.6	175.0
Kesler-Lee	61.5	−61.1	176.9
TOTAL	10.6	−0.20	44.4
Refractive Index			
API @ 20°C	0.0368	−0.0367	0.0993
Lindee-Whitter			
@ 20°C	0.0315	−0.0131	0.0303
TOTAL @ 20°C	0.0021	0.0	0.0074
TOTAL @ 60°C	0.0021	0.0	0.0074

Source: Dhulesia [1]

2.4.3 n-d-M Method

The n-d-M correlation is an ASTM (D-3238-85) method that uses refractive index (n), density (d), average molecular weight (MW), and sulfur (S) to estimate the percentage of total carbon distribution in aromatic ring structure (% C_A), naphthenic ring structure (% C_N), and paraffin chains (% C_P). Both refractive index and density are either measured or estimated at 20°C (68°F). Appendix 4 shows formulas used to calculate the carbon distribution. Note that the n-d-M method calculates, for example, the percent of carbon in aromatic ring. For instance, if there was a toluene molecule in the feed, the n-d-M method predicts six aromatic carbons (86%) versus the actual seven carbons.

ASTM D-2502 is one of the most accurate methods of determining molecular weight. The method uses viscosity measurements; in the absence of viscosity data, the molecular weight can be estimated using the TOTAL correlation.

The n-d-M method is very sensitive to both refractive index and density. It calls for measurement or estimation of the feed refractive index at 20°C (68°F). The problem is that the majority of FCC feeds are virtually solid at 20°C; consequently, the refractometer is unable to measure the refractive index at this temperature. Therefore, to use the n-d-M method, refractive index at 20°C needs to be estimated using published correlations. For this reason, n-d-M method is usually employed in conjunction with other correlations such as TOTAL. Example 2-3 can be used to illustrate the use of n-d-M correlations.

Example 2-3

Using the feed property data in Example 2-1, determine MW, C_A, C_N and C_P using the n-d-M method.

Step 1: Molecular weight determination by ASTM method.
    1. Obtain viscosity at 100°F:
        a. Plot viscosities at 130°F and 210°F.
        b. Extrapolate to 100°F, VIS = 279 SSU.
    2. Convert viscosities from SUS to centistoke:
        a. From Appendix 6, viscosity @ 100°F = 60.0 cSt.
        b. Viscosity @ 210°F = 7.37 cSt.
    3. Obtain molecular weight:
        a. From Appendix 5, H = 372 and MW = 430.

Step 2: Calculate refractive index @ 20°C from the TOTAL correlation.

$$RI(20) = 1 + 0.8447 \times (0.913)^{1.2056} \times (728.2)^{-0.0557} \times (430)^{-0.0044}$$

$$RI(20) = 1.5046$$

Step 3: Calculate n-d-M factors.

$$v = 1.51 \times (1.5105 - 1.4750) - (0.913 - 0.8510) = +0.0271$$

$$w = (0.913 - 0.851) - 1.11 \times (1.5105 - 1.4750) = +0.0226$$

Because v is positive:

$$\%C_A = 430 \times v + \frac{3660}{MW} = 11.65 + 8.51$$

$$\%C_A = 20.16$$

Because w is positive:

$$\%C_R = 820 \times 0.0226 - 3 \times 0.48 + \frac{10,000}{430}$$

$$\%C_R = 40.35$$

$$\%C_N = 40.35 - 20.16 = 20.19$$

$$\%C_P = 100 - 40.35 = 59.65$$

2.4.4 API Method

The API method is a generalized method that predicts mole fraction of paraffinic, naphthenic, or aromatic compounds for an olefin-free hydrocarbon. The development of the equations is based on dividing the hydrocarbon into two molecular ranges: heavy fractions (200 < MW < 600) and light fractions (70 < MW < 200). Appendix 7 contains API correlations applicable to the FCC feed.

Example 2-4

Use the feed property data in Example 2-1 to calculate MW, RI(20), X_A, X_N, and X_P, employing API correlations.

$$MW = 20.486 \times \exp^{(1.165 \times 10^{-4} \times 1276 - 7.787 \times 0.913 + 1.1582 \times 10^{-3} \times 0.913 \times 1276)}$$

$$\times 1276^{1.26807} \times 0.913^{4.98308}$$

$$MW = 412.9$$

$$I = 2.341 \times 10^{-2} \times e^{(6.464 \times 10^{-4} \times 1276 + 5.1444 \times 0.913 - 3.289 \times 10^{-4} \times 1276 \times 0.913)}$$

$$\times\, 1276^{-0.407} \times (0.913)^{-3.3333}$$

$$I = 0.294$$

$$RI(20) = \left(\frac{1 + 2 \times 0.294}{1 - 0.294} \right)^{1/2}$$

$$RI(20) = 1.500$$

$$VGC = \frac{0.913 - 0.24 - 0.022 \times \log(50 - 35.5)}{0.755}$$

$$VGC = 0.8575$$

$$X_A = -403.8 + 265.7 \times \left(1.5000 - \frac{0.913}{2} \right) + 161.0 \times 0.8575$$

$$X_A = 11.5 \text{ mol}\%$$

$$X_N = 246.4 - 367.0 \times \left(1.5000 - \frac{0.913}{2} \right) + 196.3 \times 0.8575$$

$$X_N = 31.8 \text{ mol}\%$$

$$X_P = 100 - (11.5 + 31.8)$$

$$X_P = 56.7 \text{ mol}\%$$

The findings from TOTAL, n-d-M, and API are summarized in Table 2-10. The comparison illustrates how sensitive the predicted feed composition is to the refractive index @ 20°C. For instance, using the

Table 2-10
Comparison of the Findings Among the 3 Correlations

	API	n-d-M	TOTAL
Refractive Index			
@ 20°C	1.5000		1.5105
Molecular Weight	413	430	446
Carbon Content:	Mol%	Wt%	Wt%
Aromatic	11.5, (14.3)*	(20.2)*,(8.8)†	19.2, (12.5)†
Naphthene	31.8, (27.9)*	(20.2)*,(41.1)†	
Paraffin	56.7, (57.8)*	(59.6)*,(50.1)†	

Uses RI(20) from n-d-M correlation to determine composition.
†*Uses RI(20) from API correlation to determine composition.*

TOTAL correlation, there is a 35% drop in the aromatic content in using RI(20) = 1.5000 instead of RI(20) = 1.5105. To use these correlations, every effort should be made to obtain accurate and consistent values for refractive index @ 20°C. Given the refractive index at any given temperature, the RI(20) can be calculated from the following equation. Example 2-5 illustrates the use of the equation.

$$RI(20) = RI(t) + 6.25 \times (t - 20) \times 10^{-4}$$

$$t = temp., °C$$

Example 2-5

Given the refractive index @ 67°C = 1.4810, determine the refractive index @ 20°C.

$$RI(20) = 1.4810 + 6.25 \times (67 - 20) \times 10^{-4}$$

$$RI(20) = 1.5104$$

(Note that the calculated RI(20) closely matches that using the TOTAL correlation.)

SUMMARY

It is important to characterize FCC feeds as to their molecular structure. Once the molecular configuration is known, kinetic models can be developed to predict product yields. The above simplified correlations do a reasonable job of defining hydrocarbon type and distribution in FCC feeds. Each correlation, within the range that it was developed, provides satisfactory results.

A clear understanding of feed physical properties is essential to successful work in the areas of troubleshooting, catalyst selection, unit optimization, and any planned revamp.

REFERENCES

1. Dhulesia, H., "New Correlations Predict FCC Feed Characterizing Parameters," *Oil & Gas Journal,* January 13, 1986, pp. 51–54.
2. ASTM, "Carbon Distribution and Structural Group Analysis of Petroleum Oils by the n-d-M Method," ASTM Standard D-3238-85, 1985.
3. Riazi, M. R., and Daubert, T. E., "Prediction of the Composition of Petroleum Fractions," *Ind. Eng. Chem. Process Dev.,* Vol. 19, No. 2, 1982, pp. 289–294.
4. ASTM, "Estimation of Molecular Weight (Relative Molecular Mass) of Petroleum Oils from Viscosity Measurements," ASTM Standard D-2502, 1982.
5. Flanders, R. L., *Proceedings of the 35th Annual NPRA Q&A Session on Refining and Petrochemical Technology,* Philadelphia, Pa., 1982, p. 59.
6. Wollaston, E. G., Forsythe, W. L., and Vasalos, I. A., "Sulfur Distribution in FCC Products," *Oil & Gas Journal,* August 2, 1971, pp. 64–69.
7. Huling, G. P., McKinney, J. D., and Readal, T. C., "Feed-Sulfur Distribution in FCC Products," *Oil & Gas Journal,* May 19, 1975, pp. 73–79.
8. Campagna, R. J., Krishna, A. S., and Yanik, S. J., "Research and Development Directed at Resid Cracking," *Oil & Gas Journal,* October 31, 1983, pp. 129–134.
9. Davison Div., W. R. Grace & Co., "Questions Frequently Asked About Cracking Catalysts, Grace Davison *Catalagram,* No. 64, 1982, p. 29.
10. Andreasson, H. U. and Upson, L. L., "What Makes Octane," presented at Katalistiks' 6th Annual FCC Symposium, Munich, Germany, May 22–23, 1985.
11. Van, K. B., Gevers, A., and Blum, A., "FCC Unit Monitoring and Technical Service," presented at 1986 Akzo Chemicals Symposium, Amsterdam, The Netherlands.

12. Scherzer, J. and McArthur, D. P., "Nitrogen Resistance of FCC Catalysts," presented at Katalistiks' 8th Annual FCC Symposium, Venice, Italy, 1986.

13. Dougan, T. J., Alkemade, V., Lakhampel, B., and Brock, L. T., "Advances in FCC Vanadium Tolerance," presented at NPRA Annual Meeting, San Antonio, Texas, March 20, 1994; reprinted in Grace Davison *Catalagram.*

CHAPTER 3

FCC Catalysts

The introduction of zeolite in commercial FCC catalysts in the early 1960s was one of the most significant advances in the history of cat cracking. It provided the refiner a greater profit with little capital investment. Simply stated, zeolite cracking catalysts were and still are the biggest bargains of all time for the refiner. Continual improvements in catalyst technology have enabled refiners to continue to meet the demands of their market with minimum capital investment.

Compared to previous amorphous silica-alumina catalysts, the zeolite catalysts are more active and more selective. The higher activity and selectivity translate to more profitable liquid product yields and additional cracking capacity. In addition, to take full advantage of the zeolite catalysts, refiners have revamped older units to economically crack more of the heavier, lower-value feedstocks.

A complete discussion of FCC catalysts would fill another book. The intent of this chapter is to provide the information needed to select the proper catalyst and to troubleshoot the unit's operation. The key topics discussed are as follows:

- Catalyst Components
- Catalyst Manufacturing Techniques
- Fresh Catalyst Properties
- Equilibrium Catalyst Analysis
- Catalyst Management
- Catalyst Evaluation
- Additives

3.1 CATALYST COMPONENTS

FCC catalysts are in the form of fine powders with an average particle size diameter in the range of 75 microns. A modern cat cracking catalyst has four major component systems: zeolite, matrix, binder, and filler.

3.1.1 Zeolite

The zeolite, or more properly, faujasite, is the key ingredient of the FCC catalyst. Its role in the catalyst is to provide product selectivity and much of the catalytic activity. The catalyst's performance depends largely on the nature and quality of the zeolite. Understanding the zeolite structure, types, cracking mechanism, and properties is essential in choosing the "right" catalyst to produce the desired yields.

Zeolite structure

Zeolite, sometimes called molecular sieve, has a well-defined lattice structure. Its basic building blocks are silica and alumina tetrahedra. Each tetrahedron (Figure 3-1) consists of a silicon or aluminum atom at the center of the tetrahedron, with oxygen atoms at the corners.

Zeolite lattices have an organized network of very small pores. The pore diameter of nearly all of today's FCC zeolite is approximately 8.0 angstroms (°A). These small openings, with an internal surface area of roughly 600 square meters per gram, do not readily admit hydrocarbon molecules with a molecular diameter greater than 8.0 °A to 10 °A.

The elementary building block of the zeolite crystal is called a unit cell. The unit cell size (UCS) is the distance between the repeating cells in the zeolite structure. One unit cell in a typical fresh Y-zeolite lattice contains 192 framework atomic positions: 55 atoms of aluminum and 137 atoms of silicon. This corresponds to a silica (SiO_2) to alumina (Al_2O_3) molal ratio (SAR) of 5. The UCS is an important parameter in characterizing the zeolite structure.

Zeolite chemistry

As stated above, a typical zeolite consists of silicon and aluminum atoms that are tetrahedrally joined by four oxygen atoms. Silicon is in a +4 oxidation state, therefore, a tetrahedron containing silicon is neutral in charge. In contrast, aluminum is in a +3 oxidation state. This indicates that each tetrahedron containing aluminum has a net charge of −1 which must be balanced by a positive ion.

Solutions containing sodium hydroxide are commonly used in synthesizing the zeolite. The sodium serves as the positive ion to balance the negative charge of aluminum tetrahedron. This zeolite is called soda Y or NaY. The NaY zeolite is not hydrothermally stable because

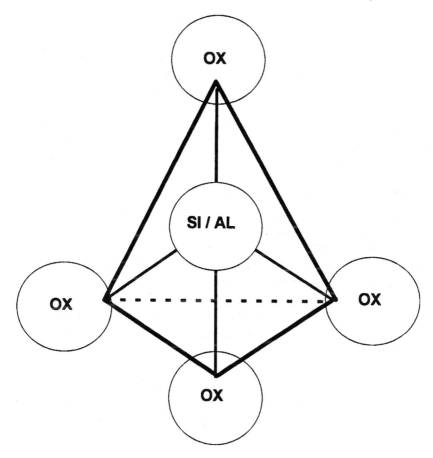

Figure 3-1.   Silicon/aluminum-oxygen tetrahedron.

of the high sodium content. Ammonium ion is frequently used to displace sodium. Upon drying the zeolite, ammonia is vaporized. The resulting acid sites are both the Bronsted and Lewis types. The Bronsted acid sites can be further exchanged with rare earth material such as cerium and lantheium to enhance their strengths. The zeolite activity comes from these acid sites.

Zeolite types

Zeolites employed in the manufacture of the FCC catalyst are synthetic versions of naturally occurring zeolites called faujasites. There are about 40 known natural zeolites and over 150 zeolites which have been synthesized. Of this number, only a few have found commercial applications. Table 3-1 shows properties of the major synthetic zeolites. The zeolites with applications to FCC are Type X, Type Y, and ZSM-5. Both X and Y zeolites have essentially the same crystalline structure. The major difference is that the X zeolite has a lower silica/alumina ratio than the Y zeolite. The X zeolite also has a lower thermal and hydrothermal stability than Y zeolite. Some of the earlier FCC zeolite catalysts contained X zeolite, however, virtually all of today's catalysts contain Y zeolite or variations thereof (Figure 3-2).

ZSM-5 is a versatile zeolite that some refiners add to the unit to increase olefin yields and octane of the FCC gasoline. Its application is further discussed in Section 3.7.3.

Until the late 1970s, the NaY zeolite was mostly ion exchanged with rare earth components. Rare earth components such as lanthanum and cerium were used to replace sodium in the crystal. The rare-earth elements, being trivalent, simply form "bridges" between two to three acid sites in the zeolite framework. The bridging basically protects acid sites from being ejected from the framework and thus stabilizes the zeolite structure. Consequently, the rare earth exchange adds to the zeolite activity and thermal and hydrothermal stability.

The reduction of lead in motor gasoline in 1986 created the need for a higher FCC gasoline octane. Catalyst manufacturers responded by adjusting the zeolite formulations, an alteration that involved expelling

Table 3-1
Properties of Major Synthetic Zeolites

Zeolite Type	Pore Size Dimensions (°A)	Silica-to-Alumina Ratio	Processes
Zeolite A	4.1	2–5	Detergent manufacturing
Faujasite	7.4	3–6	Catalytic cracking and hydrocracking
ZSM-5	5.2 × 5.8	30–200	Xylene isomerization, benzene alkylation, catalytic cracking, catalyst dewaxing, and methanol conversion.
Mordenite	6.7 × 7.0	10–12	Hydro-isomerization, dewaxing

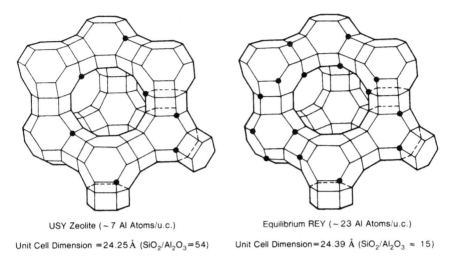

USY Zeolite (~ 7 Al Atoms/u.c.)	Equilibrium REY (~ 23 Al Atoms/u.c.)
Unit Cell Dimension =24.25 Å (SiO$_2$/Al$_2$O$_3$=54)	Unit Cell Dimension=24.39 Å (SiO$_2$/Al$_2$O$_3$ ≈ 15)

Figure 3-2. Geometry of USY and REY zeolites [14].

a number of aluminum atoms from the zeolite framework. The removal of aluminum increased SAR, reduced UCS, and in the process, lowered the sodium level of the zeolite. These changes increased the gasoline octane by raising its olefinicity. This aluminum-deficient zeolite was called *ultrastable Y,* or simply USY, because of its higher stability than the conventional Y.

Zeolite properties

The properties of the zeolite play a significant role in the overall performance of the catalyst, and familiarity with these properties increases our ability to predict catalyst response to continual changes in unit operation. From its inception in the catalyst plant, the zeolite must withstand and retain its catalytic properties under the hostile conditions of the FCC operation. The reactor/regenerator environment can cause significant changes in chemical and structural composition of the zeolite. In the regenerator, for instance, the zeolite is subjected to thermal and hydrothermal treatments. The zeolite must also retain its crystallinity against feedstock contaminants such as vanadium and sodium.

Various analytical tests can be carried out to determine zeolite properties. These tests should supply information about the strength, type, number, and distribution of acid sites. Additional tests can also provide information about surface area and pore size distribution. The three most common parameters governing zeolite behavior are as follows:

- Unit Cell Size
- Rare Earth Level
- Sodium Content

Unit Cell Size (UCS). The UCS is a measure of aluminum sites or the total potential acidity per unit cell. The negatively-charged aluminum atoms are sources of active sites in the zeolite. Silicon atoms do not possess any activity. The UCS is related to the number of aluminum atoms per cell (N_{Al}) by [1]:

$$N_{Al} + 111 \times (UCS - 24.215)$$

The number of silicon atoms (N_{Si}) is:

$$N_{Si} = 192 - N_{Al}$$

The SAR of the zeolite can be determined either from the above two equations or from a correlation such as the one shown in Figure 3-3.

The UCS is also an indicator of zeolite acidity. Because the aluminum ion is larger than the silicon ion, as the UCS decreases, the acid sites become farther apart. The strength of the acid sites is determined by the extent of their isolation from the neighboring acid sites. The close proximity of these acid sites causes destabilization of the zeolite structure. Acid distribution of the zeolite is a fundamental factor affecting zeolite activity and selectivity. Additionally, the UCS measurement can be used to indicate octane potential of the zeolite. A lower UCS presents fewer active sites per unit cell. The fewer acid sites are farther apart and therefore inhibit hydrogen transfer reactions, which in turn increases gasoline octane as well as production of C_3 and lighter components (Figure 3-4). The octane increase is due to a higher concentration of olefins in the gasoline.

Zeolites with lower UCS are initially less active than the conventional rare-earth-exchanged zeolites (Figure 3-5). However, the lower UCS zeolites tend to retain a greater fraction of their activity under severe thermal and hydrothermal treatments, hence the name ultrastable Y.

A freshly-manufactured zeolite has a relatively high UCS in the range of 24.50 °A to 24.75 °A. The thermal and hydrothermal environment of the regenerator extracts alumina from the zeolite structure

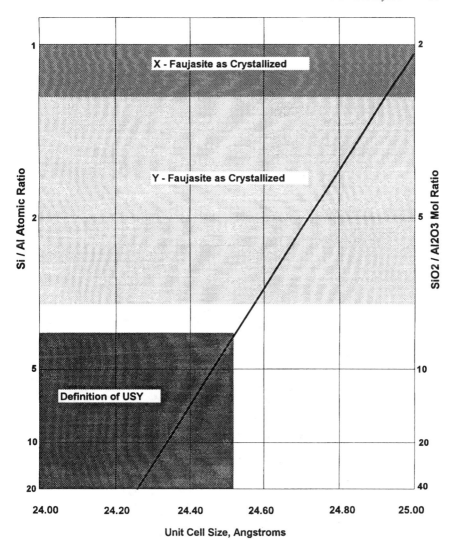

Figure 3-3. Silica-alumina ratio versus zeolite unit cell size.

and therefore reduces its UCS. The final UCS level depends on the rare earth and sodium level of the zeolite. The lower the sodium and rare earth content of the fresh zeolite, the lower UCS of the equilibrium catalyst (E-cat).

(text continued on page 88)

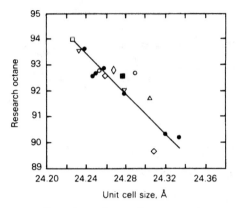

Dependence of research octane number on zeolite unit cell size

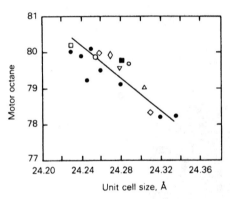

Dependence of motor octane number on zeolite unit cell size

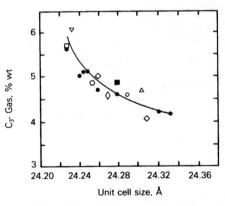

Dependence of light gas yield on zeolite unit cell size

Figure 3-4. Effects of unit cell size on octane and C_3^- gas make [4].

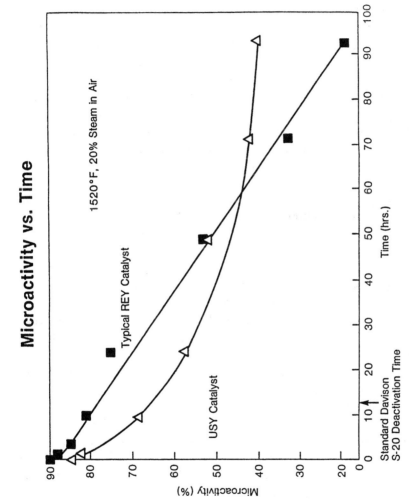

Figure 3-5. Comparison of activity retention between rare earth exchanged zeolites versus USY zeolites. *(Source: Grace Davison Octane Handbook.)*

(text continued from page 85)

Rare Earth Level. Rare earth elements serve as a "bridge" to stabilize aluminum atoms in the zeolite structure. They prevent the aluminum atoms from separating from the zeolite lattice when the catalyst is exposed to high-temperature steam in the regenerator.

A fully rare-earth-exchanged zeolite equilibrates at a high UCS, whereas a non-rare-earth zeolite equilibrates at a very low UCS in the range of 24.25 [3]. All intermediate levels of rare-earth-exchanged zeolite can be produced. The rare earth increases zeolite activity and gasoline selectivity with a loss in octane (Figure 3-6). The octane loss is due to promotion of hydrogen transfer reactions. The insertion of rare earth maintains more and closer acid sites, which promotes hydrogen transfer reactions. In addition, rare earth improves thermal and hydrothermal stability of the zeolite. To improve the activity of a USY zeolite, the catalyst suppliers frequently add some rare earth to the zeolite.

Sodium Content. The sodium on the catalyst originates either from zeolite during its manufacture or from the FCC feedstock. It is important for the fresh zeolite to contain very low amounts of sodium.

Sodium decreases the hydrothermal stability of the zeolite. It also reacts with the zeolite acid sites to reduce the catalyst activity. In the regenerator, sodium is mobile. Sodium ions tend to neutralize the strongest acid sites. In a dealuminated zeolite where the UCS is low (24.22 °A to 24.25 °A), the sodium can have an adverse affect on the gasoline octane (Figure 3-7). The loss of octane is attributed to the drop in the number of strong acid sites.

FCC catalyst vendors are now able to manufacture catalysts with a sodium content of less than 0.20 wt%. Sodium is commonly reported as the weight percent of sodium or soda (Na_2O) on the catalyst. The proper way to compare sodium is the weight fraction of sodium in the zeolite. This is because FCC catalysts have different zeolite concentrations.

UCS, rare earth, and sodium are just three of the parameters that are readily available to characterize the zeolite properties. They provide valuable information about catalyst behavior in the cat cracker. If required, additional tests can be conducted to examine other zeolite properties.

3.1.2 Matrix

The term *matrix* has different meanings to different people. For some, matrix refers to components of the catalyst other than the zeolite.

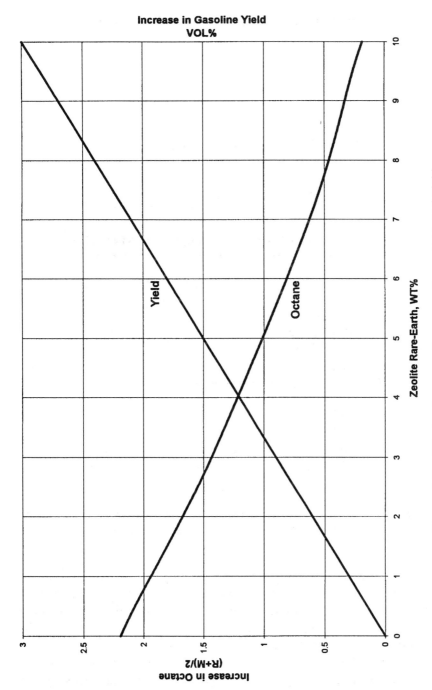

Figure 3-6. Effects of rare earth on gasoline octane and yield.

MOTOR OCTANE VS. SODIUM OXIDE

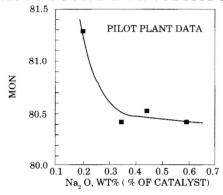

RESEARCH OCTANE VS. SODIUM OXIDE

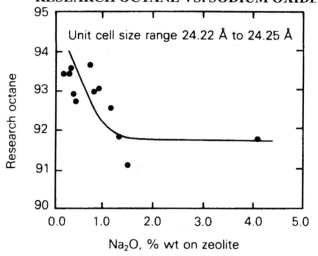

Figure 3-7. Effects of soda on motor and research octanes: motor octane vs. sodium oxide [11]; research octane vs. sodium oxide [4].

For others, matrix is a component of the catalyst aside from the zeolite having catalytic activity. Yet for others, matrix refers to the catalyst binder. In this chapter, matrix means components of the catalyst other than zeolite and the term *active matrix* means the component of the catalyst other than zeolite having catalytic activity.

Alumina is the source for an active matrix. Most active matrices used in FCC catalysts are amorphous. However, some of the catalyst suppliers incorporate a form of alumina that also has a crystalline structure.

Active matrix contributes significantly to the overall performance of the FCC catalyst. The zeolite pores are not suitable for cracking of the large hydrocarbon molecules generally having an end point >900°F; they are too small to allow diffusion of the large molecules to the cracking sites. An effective matrix must have a porous structure to allow diffusion of hydrocarbons into and out of the catalyst.

An active matrix provides the primary cracking sites. The acid sites located in the catalyst matrix are not as selective as the zeolite sites but are able to crack larger molecules that are hindered from entering the small zeolite pores. The active matrix precracks heavy feed molecules for further cracking at the internal zeolite sites. The result is a synergistic interaction between matrix and zeolite in which the activity attained by their combined effects can be greater than the sum of their individual effects [2].

An active matrix can also serve as a trap to catch some of the vanadium and basic nitrogen. The high boiling fraction of the FCC feed usually contains metals and basic nitrogen that poison the zeolite. One of the advantages of an active matrix is that it guards the zeolite from becoming deactivated prematurely by these impurities.

3.1.3 Filler and Binder

The *filler* is a clay incorporated into the catalyst to dilute its activity. Kaoline [$Al_2(OH)_2.Si_2O_5$] is the most common clay used in the FCC catalyst. One FCC catalyst manufacturer uses kaoline clay as a skeleton to grow the zeolite in situ.

The *binder* serves as a glue to hold the zeolite, the matrix, and the filler together. Binder may or may not have catalytic activity. The importance of the binder becomes more prominent with catalysts that contain high concentrations of zeolite.

The functions of the filler and the binder are to provide physical integrity (density, attrition resistance, particle size distribution, etc.), a heat transfer medium, and a fluidizing medium in which the more important and expensive zeolite component is incorporated.

In summary, zeolite is the primary ingredient for selective cracking. Changes to the zeolite will affect activity, selectivity, and product quality. An active matrix can improve bottoms cracking and resist vanadium and nitrogen attacks. But a matrix containing very small pores can suppress strippability of the spent catalyst and increase hydrogen yield in the presence of nickel. Clay and binder provide physical integrity and mechanical strength.

3.2 CATALYST MANUFACTURING TECHNIQUES

The manufacturing process of the modern FCC catalyst is divided into two general groups. Except for one manufacturer (Engelhard Corporation), the catalyst is manufactured by making zeolite and matrix independently and using a binder to hold them together. Engelhard primarily manufactures FCC catalysts by growing zeolite components within the matrix material in situ. The following sections provide a general description of zeolite synthesis.

3.2.1 Conventional Zeolite (REY, REHY, HY)

NaY zeolite is produced by digesting a mixture of silica, alumina, and caustic for several hours at a prescribed temperature until crystallization occurs (Figure 3-8). Typical sources of silica and alumina are sodium silicate and sodium aluminate. Crystallization of Y-zeolite typically takes 10 hours at about 210°F. Production of a quality zeolite requires proper control of temperature, time, and pH of the crystallization solution. NaY zeolite is separated after filtering and water-washing of the crystalline solution.

A typical NaY zeolite contains approximately 13 wt% Na_2O. To enhance activity and thermal and hydrothermal stability of NaY, the sodium level must be reduced. This is normally done by the ion exchanging of NaY with a medium containing rare-earth cations and/or hydrogen ions. Ammonium sulfate solutions are frequently employed as a source for hydrogen ion.

At this state of the catalyst synthesis, there are two approaches for further treatment of NaY. Depending on the particular catalyst and also the catalyst supplier, further treatment (rare earth exchanged) of NaY can be accomplished either before or after its incorporation into the matrix. Post-treatment of the NaY zeolite is simpler but may reduce ion exchange efficiency.

3.2.2 USY Zeolite

An ultrastable or a dealuminated zeolite (USY) is produced by replacing some of the aluminum ions in the framework with silicon. The conventional technique (Figure 3-9) includes the use of a high-temperature (1300°F–1500°F) steam calcination of HY zeolite.

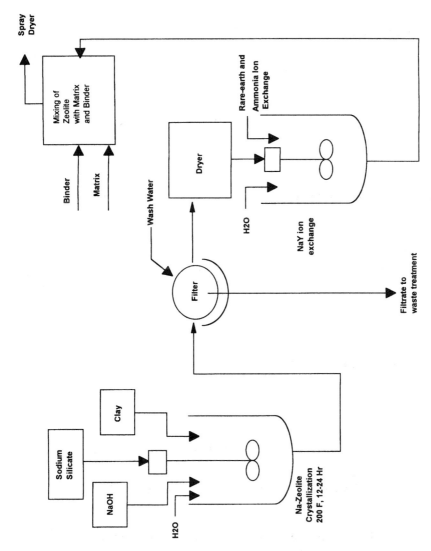

Figure 3-8. Typical manufacturing steps to produce FCC catalyst.

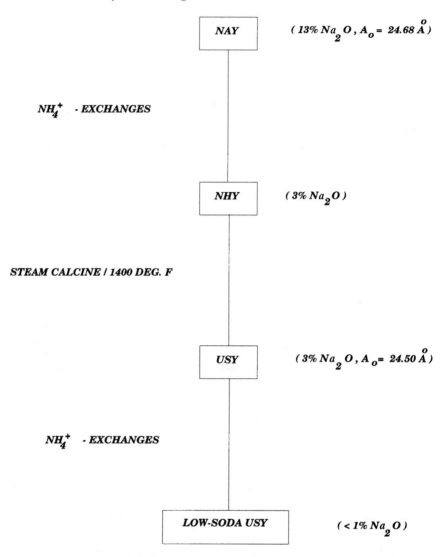

Figure 3-9. Synthesis of USY zeolite (NAY).

Acid leaching, chemical extraction, and chemical substitution are all forms of dealumination that have become popular in recent years. The main advantage of these processes over conventional dealumination is the removal of the nonframework or occluded alumina from the zeolite cage structure. A high level of occluded alumina residing in the crystal is thought to have an undesirable impact on product selectivities by yielding more light gas and LPG; however, this has not been proven commercially.

In the manufacturing of USY catalyst, the zeolite, the clay, and the binder are slurried together. If the binder is not active, an alumina component, having catalytic properties, may also be added. The well-mixed slurry solution is then fed to a spray dryer. The function of a spray dryer is to form microspheres by evaporating the slurry solution, through the use of atomizers, in the presence of hot air. The type of spray dryer and the drying conditions determine the size and distribution of catalyst particles.

3.2.3 Engelhard Process

Engelhard's FCC catalyst technology is based mainly on growing zeolite within the kaolin-based particles. The aqueous solution of kaolin is spray dried to form microspheres. The microspheres are hardened in a high-temperature (1800°F) calcination process. The NaY is produced by digestion of solutions of caustic, metakaolin, and mullite microspheres. The well-mixed solution is filtered and washed prior to ion exchange and any final treatment.

3.3 FRESH CATALYST PROPERTIES

With each shipment of fresh catalyst, the catalyst suppliers typically mail refiners an inspection report that contains data on the catalyst's physical and chemical properties. This data is valuable and should be monitored closely to ensure that the catalyst received meets the agreed specifications. A number of refiners independently analyze random samples of the fresh catalyst to confirm the reported properties. In addition, quarterly review of the fresh catalyst properties with the catalyst vendor will ensure that the control targets are being achieved. The particle size distribution (PSD), the sodium (Na), the rare earth (RE), and the surface area (SA) are some of the parameters in the inspection sheet that require close attention.

3.3.1 Particle Size Distribution (PSD)

The PSD is an indicator of the fluidization properties of the catalyst. In general, fluidization improves as the fraction of the 0-40 micron particles is increased; however, a higher percentage of 0-40 micron particles will also result in greater catalyst losses.

The fluidization characteristics of an FCC catalyst depend largely on the unit's mechanical configuration. The percentage of less than 40 microns in the circulating inventory is a function of cyclone efficiency. In units with good catalyst circulation, it may be economical to minimize the fraction of less than 40 micron particles. This is because after a few cycles, most of the 0-40 microns will escape the unit via the cyclones.

The catalyst manufacturers control PSD of the fresh catalyst mainly through the spray-drying cycle. In the spray dryer, the catalyst slurry must be atomized effectively to achieve proper distribution. As illustrated in Figure 3-10, the PSD does not have a normal distribution shape.

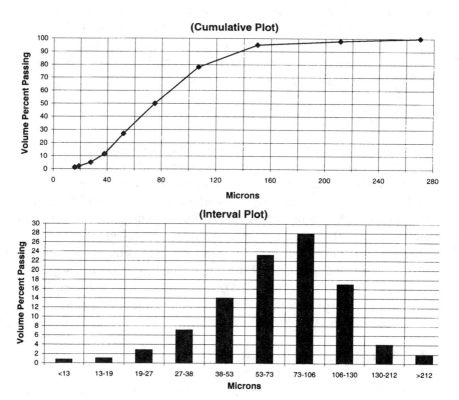

Figure 3-10. Particle size distribution of a typical FCC catalyst.

The average particle size (APS) is not actually the average size of the catalyst particles but rather the median value.

3.3.2 Surface Area (SA), M²/g

The reported surface area is the combined surface area of zeolite and matrix. In zeolite manufacturing, the measurement of the zeolite surface area is one of the procedures used by catalyst suppliers to control quality. The surface area is commonly determined by the amount of nitrogen adsorbed by the catalyst.

The surface area correlates fairly well with the fresh catalyst activity. Upon request, catalyst suppliers can also report the zeolite surface area. This data is useful in that it is proportional to the zeolite content of the catalyst.

3.3.3 Sodium (Na), wt%

Sodium plays an intrinsic part in the manufacturing of FCC catalysts. Its detrimental effects are well known, and because it deactivates the zeolite and reduces the gasoline octane, every effort should be made to minimize the amount of sodium in the fresh catalyst. The catalyst inspection sheet expresses sodium or soda (Na_2O) as the weight percent on the catalyst. When comparing different grades of catalysts, it is more practical to express the sodium content on the zeolite.

3.3.4 Rare Earth (RE), wt%

Rare earth is a generic name for 14 metallic elements of the lanthanide series. These elements have similar chemical properties and are usually supplied as a mixture of oxides extracted from ores such as bastnaesite or monazite.

Rare earth improves the catalyst activity (Figure 3-11) and hydrothermal stability. Catalysts can have a wide range of rare earth levels depending on the refiner's objectives. Similar to sodium, the inspection sheet shows rare earth (RE) or rare earth oxide (RE_2O_3) as the weight percent on the catalyst. Again, when comparing different catalysts, the concentration of RE on the zeolite should be used.

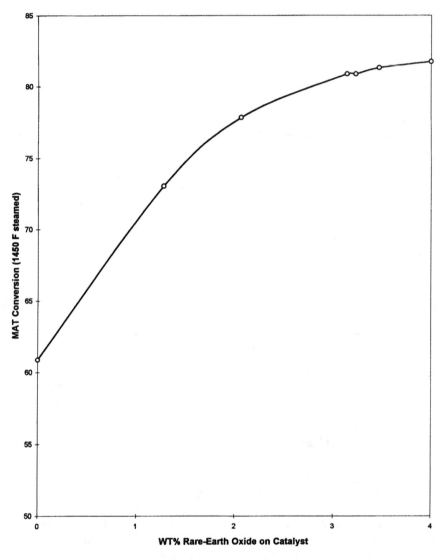

Figure 3-11. Effect of rare earth on catalyst activity.

3.4 EQUILIBRIUM CATALYST ANALYSIS

Refiners send E-cat samples to catalyst manufacturers on a regular basis. As a service to the refiners, the catalyst suppliers provide analyses of the samples in a form similar to the one shown in Figure 3-12. Although the

Date Rec'd	Sample ID	MA	Dated	C.F.	G.F.	S.A. m²/gm	P.V. cc/gm	A.B.D. gm/cc	0-20 Wt%	0-40 Wt%	0-80 Wt%	APS	Al₂O₃ Wt% D.B.	Na Wt% D.B.	Fe Wt% D.B.	C Wt% D.B.	V ppm D.B.	Ni ppm D.B.	Cu ppm D.B.	Sb ppm D.B.	RE2O3	UCS	Remarks
102992		68	102692	1.4	2.5	166	.31	.94	1	8	67	69	41.5	.31	.52	.03	1448	1272	29	579	.54	24.27	
111792		64	111792	1.4	2.5	166	.31	.95	0	9	65	70	41.0	.31	.50	.04	1321	1196	27	531	.56	24.26	MA CONF.
111992		68	111692	1.5	2.1	165	.30	.95	0	5	60	74	40.7	.30	.51	.04	1321	1206	28	551	.56		
120492		65	113092	1.5	1.4	162	.30	.94	0	6	66	70	40.7	.29	.50	.04	1452	1327	27	585	.57	24.27	SA CONF.
010593		64	122892	1.5	3.1	159	.29	.96	1	8	70	68	40.7	.30	.54	.04	1378	1326	31	585	.61		
011493		66	011193	1.4	2.9	160	.28	.95	0	8	68	69	41.1	.30	.53	.03	1637	1552	30	672	.60	24.26	MA CORR'D
012193		67	012193	1.4	3.1	158	.29	.95	0	6	67	70	41.1	.30	.54	.03	1677	1598	30	682	.58		
020493		66	020193	1.2	2.8	154	.30	.95	0	7	69	69	41.4	.30	.53	.05	1835	1594	31	709	.73	24.28	SA & MET. CONF.
022293		63	021293	1.6	2.8	157	.29	.95	0	7	69	71	41.6	.28	.53	.05	1837	1586	31	702	.75		
032293		66	031593	1.6	2.8	153	.30	.95	0	7	65	71	42.0	.28	.53	.03	1591	1255	30	519	.78		
041493		63	040593	1.6	3.2	161	.29	.96	0	7	66	70	40.4	.28	.52	.03	1500	1074	30	407	.61	24.26	
041593		61	041493	1.7	2.7	157	.31	.96	1	5	66	71	40.5	.28	.54	.04	1555	1096	29	421	.57		
042993		65	042603	1.5	3.3	152	.29	.97	1	6	64	71	41.1	.29	.52	.05	1654	1104	28	502	.67	24.29	
052193		65	051793	1.6	3.4	152	.30	.96	0	6	64	71	41.1	.31	.31	.02	1801	1138	28	485	.73	24.28	
060193		67	052493	1.4	2.5	152	.30	.96	0	7	73	70	42.3	.32	.52	.04	1803	1127	27	492	.76		
060493		65	053193	1.3	3.0	152	.30	.98	0	6	72	67	42.3	.32	.64	.03	1900	1171	27	477	.78		
061093		67	060793	1.3	3.0	155	.29	.97	0	6	64	72	41.3	.33	.55	.04	1903	1202	32	493	.61		
061793		64	061493	1.9	3.8	161	.30	.98	0	7	70	72	41.4	.28	.58	.06	1910	1216	33	488	.54		
062193		66	062193	1.4	3.5	159	.29	.98	0	6	68	68	41.8	.26	.52	.05	1976	1265	32	559	.51		
062893		68	062893	1.4	2.9	167	.29	.97	0	5	62	72	41.9	.25	.51	.03	2001	1256	32	527	.52		Z/Me-95/66MA CONF
070293		70	070593		3.5	166	.30	.97	1	5	66	70	42.3	.27	.50	.04	2040	1284	31	570	.54		MA CONF.
071393		67	071993	1.5	2.9	174	.30	.96	1	7	66	71	42.3	.31	.48	.06	2262	1393	30	589	.65	24.27	
072393		69	080293	1.5	2.9	182	.32	.96	0	6	66	71	41.5	.31	.51	.03	2020	1249	30	533	.64		MA CONF.
080693		64	081693	1.7	3.1	180	.32	.95	1	6	66	69	42.0	.31	.50	.05	1747	1091	32	447	.53	24.27	SA CORR'D
081693		68	082793	1.7	2.8	166	.30	.96	0	6	72	68	41.7	.30	.48	.05	1722	1067	31	472	.60		PV CORR'D
082793		68	083093	1.7	3.4	168	.30	.96	1	8	73	68	41.9	.31	.53	.04	1631	1001	29	432	.68	24.28	SA CORR'D
090393		68	092793	1.6	3.9	161	.30	.95	1	7	82	71	42.0	.32	.52	.03	1663	1094	32	435	.57	24.30	SA CONF.
093093		67	110193	1.8	3.7	161	.29	.98	0	9	71	67	41.3	.33	.54	.04	1800	1178	29	320	.64	24.29	
110493		67	111593	1.8	2.7	178	.32	.92	0	4	64	69	41.5	.33	.52	.04	1994	1317	33	484	.59	24.25	SA AVG. OF 2
111993		67	122793	1.6	2.9	186	.33	.93	0	4	68	72	41.2	.32	.49	.05	1705	1077	24	411	.55	24.28	SA CONF.
123093		67	011094	1.6	2.6	179	.31	.94	0	3	64	72	41.1	.32	.49	.04	1690	1085	26	427	.56	24.27	MA CONF.
011494		67	041894	1.4	2.9	182	.31	.95	1	4	64	71	41.0	.37	.49	.06	1730	1103	28	479	.62	24.27	SA CONF.
042294		67	042594	1.7	2.0	182	.31	.97	0	6	64	69	41.1	.37	.49	.06	1951	1096	26	510	.66		
042994		69	050294	1.5	3.6	183	.30	.97	0	6	58	75	40.9	.39	.50	.04	1904	1207	28	551	.67		
050994		66	052394														1921	1177	26	480	.70		
052394			061394														1847	1181	27	468	.71		
052694			062094														1645	1120	27	450	.72		
061394																		960		234			

Figure 3-12. Typical E-cat analysis.

absolute E-cat results may differ from one vendor to another, the results are most useful as a trend indicator.

The tests performed on E-cat samples provide refiners with valuable information on unit conditions. The data can be used to pinpoint potential operational, mechanical and catalyst problems because the physical and chemical properties of the E-cat provide clues on the environment to which it been exposed.

The following discussion describes each test briefly and examines the significance of these data to the refiner. The E-cat results are divided into catalytic properties, physical properties, and chemical analysis.

3.4.1 Catalytic Properties

The activity, coke, and gas factors are the tests that reflect the relative catalytic behavior of the catalyst.

Conversion (activity)

The first step in E-cat testing is to burn the carbon off the sample. The sample is then placed in a MAT unit (Figure 3-13), the heart of which is a fixed bed reactor. A certain amount of a standard gas oil feedstock is injected into the hot bed of catalyst. The activity is reported as the conversion to –430°F material. The feedstock's quality, reactor temperature, catalyst-to-oil ratio, and space velocity are four variables affecting MAT results. Each catalyst vendor uses slightly different operating variables to conduct microactivity testing, as indicated in Table 3-2.

In commercial operations, catalyst activity is affected by operating conditions, feedstock quality, and catalyst characteristics. The MAT separates catalyst effects from feed and process changes. Feed contaminants, such as vanadium and sodium, reduce catalyst activity. E-cat activity is also affected by fresh catalyst makeup rate and regenerator conditions.

Coke factor (CF), gas factor (GF)

The CF and GF represent the coke- and gas-forming tendencies of an E-cat compared to a standard steam-aged catalyst sample at the same conversion. The CF and GF are influenced by the type of fresh catalyst and the level of metals deposited on the E-cat. Both the coke

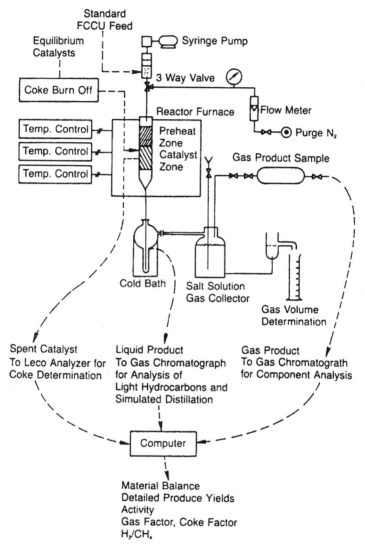

Figure 3-13. Typical MAT equipment [3].

and gas factors can be indicative of the dehydrogenation activity of the metals on the catalyst. The addition of amorphous alumina to the catalyst will tend to increase the nonselective cracking which forms coke and gas.

Table 3-2
Equilibrium Microactivity Test Conditions

Tester (United States)	Temp. °F	Cat-to-Oil Wt. Ratio	WHSV hr⁻¹	Catalyst Contact Time, Seconds	Feed Source	Reactor Type
AKZO	900	3.0	16	75	(1)	Isothermal
Davison	980	4.0	30	30	(2)	Isothermal
Engelhard	910	5.0	15	48	(3)	Isothermal

(1) Light Arabian
(2) Sour Import Heavy Gas Oil
(3) Mid-Continent

MAT Gas Oil Properties

Properties	AKZO*	Davison†	Engelhard**
°API Gravity	28.1	22.5	28.6
IBP, °F	350	423	373
50%, °F	730	755	732
90%, °F	948	932	899
Concarbon, wt%	0.04	0.25	0.22
Sulfur, wt%		2.59	0.52
Total Nitrogen, ppmv	1044	860	675
Aromatics, vol%		21.7	30
Naphthenes, vol%		19.6	28
Paraffins		58.7	42

*ASTM standard MAT feed (unavailable).
†Grace Davison Catalagram, No. 79, 1989
**Engelhard Catalyst Report, No. TI-825

3.4.2 Physical Properties

The tests that reflect physical properties of the catalyst involve surface area, average bulk density, pore volume, and particle size distribution.

Surface area (SA), M²/g

For an identical fresh catalyst, the surface area of an E-cat is an indirect measurement of its activity. The SA is the sum of zeolite and matrix surface areas. Hydrothermal conditions in the cat cracker destroy the zeolite cage structure and thus reduce its surface area. They also dealuminate the zeolite framework. Hydrothermal treatment has less effect on the matrix surface area, but the matrix surface area is affected by the collapse of small pores to become larger pores.

Apparent bulk density (ABD), g/cc

Bulk density can be used to troubleshoot catalyst flow problems. A too-high ABD can restrict fluidization, and a too-low ABD can result in excessive catalyst loss. Normally, the ABD of the equilibrium catalyst is higher than the fresh catalyst ABD due to thermal and hydrothermal changes in pore structure that occur in the unit.

Pore volume (PV), cc/g

Pore volume is an indication of the quantity of voids in the catalyst particles and can be a clue in detecting the type of catalyst deactivation that takes place in a commercial unit. Hydrothermal deactivation has very little effect on pore volume, whereas thermal deactivation decreases pore volume.

Pore diameter (°A)

The average pore diameter (APD) of a catalyst can be calculated from the E-cat analysis sheet by using the following equation:

$$APD\ (^\circ A) = \frac{PV \times 4 \times 10,000}{SA}$$

Example 3-1

For an E-cat with a PV = 0.40 cc/g and SA = 120 m²/g, determine APD.

APD = 133 °A

Particle size distribution (PSD)

PSD is an important indicator of the fluidization characteristics of the catalyst, cyclone performance, and the attrition resistance of the catalyst. A drop in fines content indicates the loss of cyclone efficiency.

This can be confirmed by the particle size of fines collected downstream of the cyclones. An increase in fines content of the E-cat indicates increased catalyst attrition. This can be due to changes in fresh catalyst binder quality, steam leaks, and/or internal mechanical problems such as those involving the air distributor or slide valves.

3.4.3 Chemical Properties

The key elements that characterize chemical composition of the catalyst are alumina, sodium, metals, and carbon on the regenerated catalyst.

Alumina (Al_2O_3)

The alumina content of the E-cat is the total weight percent of alumina (active and inactive) in the bulk catalyst. The alumina content of the E-cat is directly related to the alumina content of the fresh catalyst. When changing catalyst grades, the alumina level of the E-cat is often used to determine the percent of new catalyst in the unit.

Sodium (Na)

The sodium in the E-cat is the sum of sodium added with the feed and sodium on the fresh catalyst. A number of catalyst suppliers report sodium as soda (Na_2O). Sodium deactivates the catalyst acid sites and causes collapse of the zeolite crystal structure. Sodium can also reduce the gasoline octane, as discussed earlier.

Nickel (Ni), vanadium (V), iron (Fe), copper (Cu)

These metals, when deposited on the E-cat catalyst, increase coke- and gas-making tendencies of the catalyst. They cause dehydrogenation reactions, which increase hydrogen production and decrease gasoline yields. Vanadium can also destroy the zeolite activity and thus lead to lower conversion. The deleterious effects of these metals also depend on the regenerator temperature: The rate of deactivation of a metal-laden catalyst increases as regenerator temperature increases.

These contaminants originate largely from the heavy (1050+°F), high-molecular weight fraction of the FCC feed. The quantity of these metals on the E-cat is determined by their levels in the feedstock and the catalyst addition rate. Essentially, all these metals in the feed are deposited on the catalyst. Most of the iron on the E-cat comes from metal scale from piping and from the fresh catalyst.

Metals content of the equilibrium catalyst can be determined fairly accurately by conducting a metals balance around the unit.

$$\text{Metals}_{in} - \text{Metals}_{out} = \text{Metals Accumulated}$$

This is a first order differential equation. Its solution is:

$$M_e = A + [M_0 - A] \times e^{\frac{(C_a \times t)}{I}}$$

Where:

M_e = E-cat Metals Content, ppm
A = $(W \times M_f)/C_a$
W = Feed rate, lb/day
M_f = Feed Metals, ppm
C_a = Catalyst Addition Rate, lb/day
M_0 = Initial Metals on the E-cat, ppm
t = Time, day
I = Catalyst Inventory, lb

At steady state, the concentration of any metal on catalyst is:

$$M_e = A = \frac{W \times M_f}{C_a}$$

$$M_e = \frac{\frac{141.5}{131.5 + °API_{feed}} \times 350.4 \times M_f}{B}$$

B = Catalyst Addition Rate, pounds of catalyst per barrel of feed

Figure 3-14 is the graphical solution to the above equation and can be employed to estimate metals content of the E-cat based on feed metals and catalyst addition rate.

Carbon (C)

The deposition of carbon on the E-cat during cracking will temporarily block some of the catalytic sites. The carbon, or more accurately the coke, on the regenerated catalyst (CRC) will lower the catalyst activity and, therefore, the conversion of the feed to valuable products (Figure 3-15).

The CRC is an important parameter for a unit operator to monitor periodically. Most FCC units check for CRC on their own, usually daily. The CRC is an indicator of regenerator performance. If the CRC shows signs of increasing, this could reveal malfunction of the regenerator's air/spent catalyst distributors. It should be noted that the MAT numbers reported on the E-cat sheet are determined after the CRC has been completely burned off.

3.5 CATALYST MANAGEMENT

Depending on the design of a cat cracker, the circulating inventory can contain 30–1,200 tons of catalyst. Fresh catalyst is added to the unit continually to replace the catalyst lost by attrition and to maintain catalyst activity. The daily makeup rate is typically 1% to 2% of inventory or 0.1 to 0.3 pounds of catalyst per barrel of fresh feed. In cases where the makeup rate for activity maintenance exceeds catalyst losses, part of the inventory is periodically withdrawn from the unit to control the catalyst level in the regenerator. Catalyst fines leave the unit with the regenerator flue gas and the reactor vapor.

The catalyst ages in the unit, losing its activity and selectivity. The deactivation in a given unit is largely a function of the unit's mechanical configuration, its operating condition, the type of fresh catalyst used, and the feed quality. The primary criterion for adding fresh catalyst is to arrive at an optimum E-cat activity level. A too-high E-cat activity will increase delta coke on the catalyst, resulting in a higher regenerator temperature. The higher regenerator temperature reduces catalyst circulation rate, which tends to offset the activity increase.

The amount of fresh catalyst added is usually a balance between catalyst cost and desired activity. Most refiners monitor the MAT data

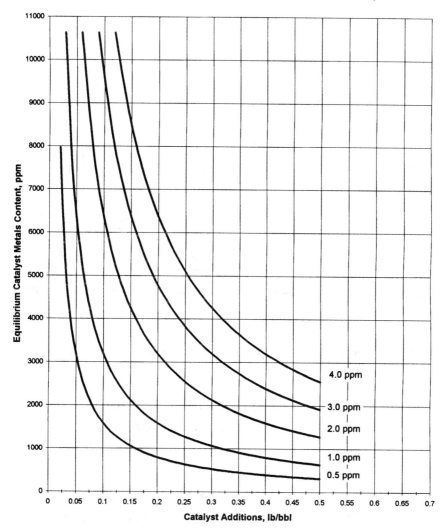

Figure 3-14. Catalyst metals content versus catalyst addition rate.

from the catalyst vendor's equilibrium data sheet to adjust the fresh catalyst addition rate. It should be noted that MAT numbers are based on a fixed-bed reactor system and, therefore, do not truly reflect the dynamics of an FCC unit. A catalyst with a high MAT number may or may not produce the desired yields. An alternate method of measuring

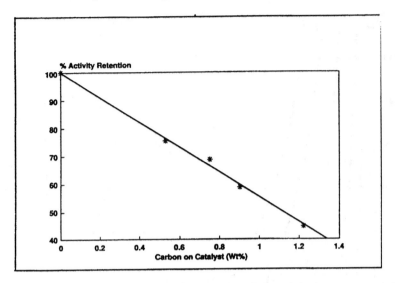

Figure 3-15. Effect of coke on the regenerated catalyst versus catalyst activity [12].

catalyst performance is *dynamic activity.* Dynamic activity is calculated as shown below:

$$\text{Dynamic Activity} = \frac{(\text{Second Order Conversion})}{(\text{Coke Yield, Wt\% of Feed})}$$

Where:

$$\text{Second Order Conversion} = \frac{(\text{MAT Conversion, Vol\%})}{(100 - \text{MAT Conversion, Vol\%})}$$

For example, a catalyst with a MAT number of 70 vol% and a 3.0 wt% coke yield will have a dynamic activity of 0.78. However, another catalyst with a MAT conversion of 68 vol% and 2.5 wt% coke yield will have a dynamic activity of 0.85. This could indicate that in a commercial unit the 68 MAT catalyst could outperform the 70 MAT catalyst due to its higher dynamic activity. Some catalyst vendors have begun reporting dynamic activity data as part of their E-cat inspection reports. The reported dynamic activity data can vary significantly from one test to another mainly due to the differences in feedstock quality between MAT and actual commercial application. In addition, the coke yield as calculated by the MAT procedure is not very accurate, and small changes in this calculation can affect the dynamic activity appreciably.

The most widely accepted model to predict equilibrium catalyst activity is based on a first-order decay type [7]:

$$A_{(t)} = A_{(0)} \times e^{-(S+K)t} + \frac{A_0 \times S}{S + K} \times \left(1 - e^{-(K+S)t}\right)$$

At steady state, the above equation reduces to:

$$A_\infty = \frac{A_0 \times S}{S + K}$$

Where:

$A_{(t)}$ = Catalyst microactivity at any time
A_0 = Catalyst microactivity at starting time
t = Time after changing catalyst or makeup rate
S = Daily Fractional Replacement Rate = Addition rate/Inventory
K = Deactivation Constant = ln $(A_t - A_0)/{-t}$

Example 3-2 below illustrates the use of the above equations.

Example 3-2

Assume a 50,000 bpd cat cracker with:

- Catalyst inventory of 300 tons
- Makeup rate of 4.0 tons per day
- Fresh catalyst MAT conversion 80 vol%
- E-cat MAT number 71.5 vol%

Determine:
a. New E-cat MAT conversion if the addition rate is reduced to 3.0 tons/day:

$$\text{a: } S = \frac{4.0}{300} = 0.01333 \text{ days}^{-1}$$

t = 300/4 = 75 days

$$\text{Deactivation Constant} = K = \frac{\left[\ln(71.5) - \ln(80)\right]}{-75.0} = 0.001498 \text{ days}^{-1}$$

$$\text{New Fractional Replacement} = \frac{3.0}{300} = 0.01 \text{ days}^{-1}$$

$$\text{The revised E-cat} = \frac{80 \times 0.01}{0.01333 + 0.001498} = 69.5 \text{ vol}\%$$

b. The new E-cat MAT conversion if the fresh catalyst MAT number is reduced from 80 vol% to 75 vol%:

$$A = \frac{75 \times 0.01333}{0.01333 + 0.001498} = 67 \text{ vol}\%$$

When a refiner changes the FCC catalyst, it is often necessary to determine the percent of the new catalyst in the unit. The following equation, which is based on a probability function, can be used to estimate the % changeover.

$$P = 1 - e^{-f \times S \times t}$$

Where:

P = Fractional Changeover
f = Retention Factor, usually in the range of 0.6 to 0.9
S = Replacement Rate = Addition rate/Inventory
t = Time, day

Example 3-3

The 300-ton inventory unit in example 3-2 is changing catalyst type and planning to add 3.5 tons per day of new catalyst. Determine the % of changeover after 60 days of operation. Assume a retention factor of 0.7.

$$P = 1 - e^{-0.7 \times 0.0117 \times 60}$$

P = 0.387 or 38.7%

Another method of calculating the % changeover is by the use of *alumina balance.*

Example 3-4

The example below illustrates the application of the alumina balance method.

For the same 300-ton inventory unit, assume the alumina (Al_2O_3) contents of the present and new fresh catalysts are 48 wt% and 38 wt%, respectively. Sixty (60) days after the catalyst switch, the alumina content of E-cat is 43 wt%. Determine % changeover:

$$\text{Fractional Changeover} = 1 - \frac{Al_2O_3(\text{new}) - Al_2O_3(\text{equil.})}{Al_2O_3(\text{new}) - Al_2O_3(\text{old})}$$

$$\text{Fractional Changeover} = 1 - \frac{38 - 43}{38 - 48} = 0.5 \equiv \rightarrow > 50\%$$

This method can also be used to calculate the catalyst retention factor. The above equations assume steady-state operation, constant unit inventory, and constant addition and loss rate.

3.6 CATALYST EVALUATION

Catalyst management is a very important aspect of the FCC process. Selection and management of the catalyst, as well as how the unit is operated, are largely responsible for achieving the desired products. Proper choice of a catalyst will go a long way toward achieving a successful cat cracker operation. Catalyst changeout is a relatively simple process and allows a refiner to select the catalyst that maximizes the profit margin. Although catalyst changeout is physically simple, it requires a lot of homework as discussed later in this section.

As many catalyst formulations are available, catalyst evaluation should be an ongoing process; however, it is not an easy task to evaluate the performance of an FCC catalyst in a commercial unit because of continual changes in feedstocks and operating conditions in addition to inaccuracies in measurements. Because of these limitations, refiners sometimes switch catalysts without identifying the objectives and limitations of their cat crackers. To assure that a proper catalyst is selected, each refiner should establish a methodology that allows identification of "real" objectives and constraints and ensures that the choice of the catalyst is based on well-thought-out technical and business merits. In today's market, there are over 120 different formulations of FCC catalysts. Refiners should evaluate catalysts mainly to maximize profit opportunity and to minimize risk. The "right" catalyst for one refiner may not necessarily be "right" for another.

A comprehensive catalyst selection methodology will have the following elements:

1. Optimize unit operation with current catalyst and vendor.
   a. Conduct test run.
   b. Incorporate the test run results into an FCC kinetic model.
   c. Identify opportunities for operational improvements.
   d. Identify unit's constraints.
   e. Optimize incumbent catalyst with vendor.
2. Issue technical inquiry to catalyst vendors.
   a. Provide Test run results.
   b. Provide E-cat sample.
   c. Provide Processing objectives.
   d. Provide Unit limitations.
3. Obtain vendor responses.
   a. Obtain catalyst recommendation.
   b. Obtain alternate recommendation.
   c. Obtain comparative yield projection.
4. Obtain current product price projections.
   a. For present and future four quarters.
5. Perform economic evaluations on vendor yields.
   a. Select catalysts for MAT evaluations.
6. Conduct MAT of selected list.
   a. Perform physical and chemical analyses.
   b. Determine steam deactivation conditions.
   c. Deactivate incumbent fresh catalyst to match incumbent E-cat.
   d. Use same deactivation steps for each candidate catalyst.

7. Perform economic analysis of alternatives.
   a. Estimate commercial yield from MAT evaluations.
8. Request commercial proposals.
   a. Consult at least two vendors.
   b. Obtain references.
   c. Check references.
9. Test the selected catalysts in a pilot plant.
   a. Calibrate the pilot plant steaming conditions using incumbent E-cat.
   b. Deactivate the incumbent and other candidate catalysts.
   c. Collect at least two or three data points on each by varying catalyst-to-oil ratio.
10. Evaluate pilot plant results.
   a. Translate the pilot plant data.
   b. Use the kinetic model to heat-balance the data.
   c. Identify limitations and constraints.
11. Make the catalyst selection.
   a. Perform economic evaluation.
   b. Consider intangibles—research, quality control, price, steady supply, manufacturing location.
   c. Make the recommendations.
12. Post selection.
   a. Monitoring transition—% changeover.
   b. Post transition test run.
   c. Confirm computer model.
13. Issue the final report.
   a. Analyze benefits.
   b. Evaluate selection methodology.

There is a redundancy of flexibility in the design of FCC catalysts. Variation in the amount and type of zeolite as well as the type of active matrix provide a great deal of catalyst options that the refiner can employ to fit its needs. For smaller refiners, it may not be practical to employ pilot plant facilities to evaluate different catalysts. In that case, the above methodology can still be used with emphasis shifted toward using the MAT data to compare the candidate catalysts. It is important that MAT data are properly corrected for temperature, "soaking time," and catalyst strippability effects.

3.7 ADDITIVES

For many years, cat cracker operators have used additive compounds for enhancing cat cracker performance. The main benefits of these additives (catalyst and feed additives) are to alter the FCC yields and to reduce the amount of pollutants emitted from the regenerator. The additives discussed in this section are: CO promoter, SO_x reduction, ZSM-5, and antimony.

3.7.1 CO Promoter

The CO promoter is added to most FCC units to assist in the combustion of CO to CO_2 in the regenerator. The promoter is added to accelerate the CO combustion in the dense phase and to minimize the high temperature excursions that occur as a result of afterburning in the dilute phase. The promoter allows uniform burning of coke, particularly if there is uneven distribution between spent catalyst and combustion air.

Regenerators operating in full or partial combustion can utilize the benefits of the CO promoter. The addition of the promoter tends to increase regenerator temperature, and therefore, the metallurgy of the regenerator internals should be checked for tolerance of the higher temperature.

The active ingredients of the promoter are the platinum group metals. The platinum, in the concentration of 300 ppm to 800 ppm, is typically dispersed on a support. The effectiveness of the promoter depends largely on its activity and stability.

Promoter is frequently added to the regenerator two to three times a day, normally at a rate of 3 to 5 pounds of promoter per ton of fresh catalyst. The use of CO promoter, particularly during unit start-ups, improves the stability of the regeneration operation. However, not every cat cracker can justify combustion-promoted operation. Heat balance, availability of combustion air, metallurgical limits, and a presence of CO boiler are some of the factors that need to be considered before using combustion promotor.

3.7.2 SO_x Additive

The coke on the spent catalyst entering the regenerator contains sulfur. In the regenerator, the sulfur in the coke is converted to SO_2 and SO_3.

The mixture of SO_2 and SO_3 is commonly referred to as SO_x, and approximately 80% to 90% of SO_x is SO_2, with the rest being SO_3. The SO_x leaves the regenerator with the flue gas and eventually is discharged to the atmosphere. Coke yield, sulfur content of the coke, regenerator operating condition, and the type of FCC catalyst are the major factors affecting SO_x emission.

The environmental impact of SO_x emissions has gained much attention over the past ten years. Presently, there is no federally mandated standard. The United States Environmental Protection Agency (EPA) has recently proposed a limit on SO_x emissions from new, modified, and reconstructed FCC units to 9.8 kilograms (22 pounds) of SO_2 per 1,000 kilograms (2,207 pounds) of coke burn-off. The Southern California Air Quality Management District (SCAQMD) board has established a limit of 60 kilograms of SO_x per 1,000 barrels of feed for the existing FCC units.

There are three common methods for SO_x abatement. These are flue gas scrubbing, feedstock desulfurization, and SO_x additive. The SO_x additive is the least costly alternative, and this is the approach practiced by many refiners.

The SO_x agent, usually a metal oxide, is added directly to the catalyst inventory. The additive works by adsorbing and chemically bonding with SO_3 in the regenerator. This stable sulfate species is carried with the circulating catalyst to the riser, where it is reduced or "regenerated" by hydrogen or water to yield H_2S and metal oxide. Table 3-3 shows postulated chemistry of SO_x reduction by an SO_x agent.

To achieve the highest efficiency of SO_x additive, it is important that:

- Excess oxygen be available; oxygen promotes the SO_2 to SO_3 reaction.
- The regenerator temperature be lower; lower temperature favors $SO_2 + 1/2\ O_2 \rightarrow SO_3$.
- The capturing agent be physically compatible with the FCC catalyst and easily regenerable in the riser.
- CO promoter be used, which oxidizes SO_2 to SO_3.
- There be a uniform distribution of air and spent catalyst.
- Operation of the reactor stripper be efficient.

Since most of the regenerators operating in full combustion mode usually operate with 1% to 3% excess oxygen, the capturing efficiency of SO_x additive is greater in full combustion than in partial combustion units.

Table 3-3
Mechanism of Catalytic SO$_x$ Reduction

A. In the Regenerator		
Sulfur in Coke (S) + O$_2$	$\rightarrow$	SO$_2$ + SO$_3$
SO$_2$ + 1/2 O$_2$	$\rightarrow$	SO$_3$
M$_x$O + SO$_3$	$\rightarrow$	M$_x$SO$_4$
B. In the Riser		
M$_x$SO$_4$ + 4H$_2$	$\rightarrow$	M$_x$S + 4H$_2$O
M$_x$SO$_4$ + 4H$_2$	$\rightarrow$	M$_x$O + H$_2$S + 3 H$_2$O
C. In the Stripper		
M$_x$S + H$_2$O	$\rightarrow$	M$_x$O + H$_2$O

Source: Thiel [9]

3.7.3 ZSM-5

ZSM-5 is Mobil Oil's proprietary shape-selective zeolite that has a different pore structure than that of Y-zeolite. The pore size of ZSM-5 is smaller than that of Y-zeolite (5.1 °A to 5.6 °A versus 8 °A to 9 °A). In addition, the pore arrangement of ZSM-5 is different than Y-zeolite as shown in Figure 3-16. The shape selectivity of ZSM-5 allows preferential cracking of long-chain, low-octane normal paraffins as well as some olefins in the gasoline fraction.

ZSM-5 additive is added to the unit to boost gasoline octane and to increase light olefin yields. ZSM-5 accomplishes this by upgrading low-octane components in the gasoline boiling range (C$_7$ to C$_{10}$) into light olefins (C$_3$, C$_4$, C$_5$). ZSM-5 inhibits paraffin hydrogenation by cracking the C$_7$+ olefins.

ZSM-5 effectiveness depends on several variables. The cat crackers that process highly paraffinic feedstock and have lower base octane will receive the greatest benefits of using ZSM-5. ZSM-5 will have little effect on improving gasoline octane in units that process naphthenic feedstock or operate at a high conversion level.

When using ZSM-5, there is almost an even trade-off between FCC gasoline volume and LPG yield. For a one-number increase in the research octane of FCC gasoline, there is a 1 to 1.5 vol% decrease in

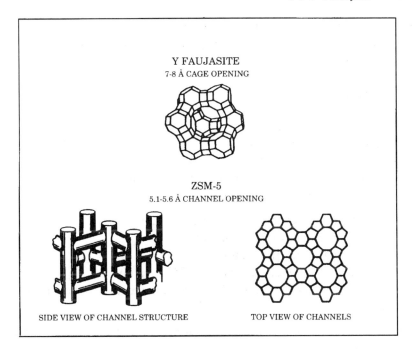

Figure 3-16. Comparison of Y faujasite and ZSM-5 zeolites [13].

the gasoline and almost a corresponding increase in the LPG. This again depends on feed quality, operating parameters, and base octane.

The decision to add ZSM-5 depends on the objectives and constraints of the unit. ZSM-5 application will increase load on the wet gas compressor, FCC gas plant, and other downstream units. Most refiners who add ZSM-5 do it on a seasonal basis, again depending on their octane need and unit limitations.

The concentration of the ZSM-5 additive should be greater than 1 percent of catalyst inventory to see a noticeable increase in the octane. An octane boost of 1 RON will typically require 2% to 5% ZSM-5 additive in the inventory. It should be noted that the proper way of quoting percentage should be by ZSM-5 concentration rather than the total additive, because the activity and attrition rate can vary from one supplier to another supplier. There are new generations of ZSM-5 additives which have nearly twice the activity of the earlier additives.

In summary, ZSM-5 provides the refiner the flexibility to increase gasoline octane and light olefins. With the introduction of reformulated gasoline, ZSM-5 could play an important role in producing iso-butylene, used as the feedstock for production of methyl tertiary butyl ether (MTBE).

3.7.4 Metal Passivation

As discussed in Chapter 2, nickel, vanadium, and sodium are the metal compounds usually present in the FCC feedstock. These metals deposit in the catalyst and thus poison the catalyst active sites. Some of the options available to refiners for reducing the effect of metals on catalyst activity are as follows:

- Increasing the fresh catalyst makeup rate.
- Using outside equilibrium catalyst.
- Employing metal passivators.
- Incorporating metal trap into the FCC catalyst.
- Using demetalizing technology to remove the metals from the catalyst.

Metal passivation in general and antimony in particular are discussed in the following section.

In recent years, several methods have been patented for chemical passivation of nickel and vanadium. Only some of the tin compounds have had limited commercial success in passivating vanadium. Although tin has been used by some refiners, it has not been proven or as widely accepted as antimony. In the case of nickel, antimony-based compounds have been most effective in reducing the detrimental effects of nickel poisoning. It should be noted that although the existing antimony-based technology is the most effective method of reducing deleterious effects of nickel, the antimony is fugitive and can be considered hazardous. In that case a bismuth-based passivator may be a better choice.

Antimony

Antimony-based passivation was introduced by Phillips Petroleum in 1976 to passivate nickel compounds in the FCC feed. Antimony is injected into the fresh feed usually with the help of a carrier such as light cycle oil. If there are feed preheaters in the unit, antimony should be injected downstream of the preheater to avoid thermal decomposition of the antimony solution in the heater tubes.

The effects of antimony passivation are usually immediate. By forming an alloy with nickel, the dehydrogenation reactions that are caused by nickel are reduced by 40% to 60%. This is evidenced by a sharp decline in dry gas and hydrogen yield.

Nickel passivation is generally economically attractive when the nickel content of the equilibrium catalyst is greater than 1200 ppm. The antimony solution should be added in proportion to the amount of nickel present in the feed. The optimum dosage normally corresponds to an antimony-to-nickel ratio of 0.3 to 0.5 on the equilibrium catalyst. Antimony's lay-down efficiency on the catalyst is in the range of 75% to 85% without recycling of slurry oil to the riser. If slurry recycle is being practiced, the lay-down efficiency is usually greater than 90%. Any antimony not deposited on the circulating catalyst ends up in the decanted oil and the catalyst fines from the regenerator.

SUMMARY

The introduction of zeolite into the FCC catalyst in the early 1960s was one of the most significant developments in the field of cat cracking. The zeolite greatly improved selectivity of the catalyst resulting in higher gasoline yields and indirectly allowing refiners to process more feed to the unit. With the introduction of reformulated gasoline, new formulations in FCC catalysts will again help refiners to meet new requirements in gasoline quality.

Since there are over 120 different FCC catalyst formulations in the market today, it is important that the refinery personal involved in cat cracker operations have some fundamental understanding of catalyst technology. This knowledge is useful in areas such as proper trouble-shooting and customizing a catalyst that would match the refiner's needs. The additive technology will be expanding in coming years. The need to produce reformulated gasoline will increase demand for the shape-selective zeolite, such as ZSM-5. The pressure from environmental agencies to reduce SO_x and NO_x will further increase the demand for additives that reduce emissions.

REFERENCES

1. Breck, D. W., *Zeolite Molecular Sieves: Structure, Chemistry, and Use.* New York: Wiley Interscience, 1974.

2. Hayward, C. M. and Winkler, W. S., "FCC: Matrix/Zeolite," *Hydrocarbon Processing,* February 1990, pp. 55–56.

3. Upson, L. L., "What FCC Catalyst Tests Show," *Hydrocarbon Processing,* November 1981, pp. 253–258.

4. Pine, L. A., Maher, P. J., and Wachter, W. A., "Prediction of Cracking Catalyst Behavior by a Zeolite Unit Cell Size Model," *Journal of Catalysis,* No. 85, 1984, pp. 466–476.

5. Magnusson, J. and Pudas, R., "Activity and Product Distribution Characteristics of the Currently Used FCC Catalyst Systems," presented at Katalistiks' 6th Annual FCC Symposium, Munich, Germany, May 22–23, 1985.

6. John G. S. and Mikovsky, R. J., "Calculation of the Average Activity of Cracking Catalysts," *Chemical Engineering Science,* Vol. 15, 1961, pp. 172–175.

7. Gaughan, J. R., "Effect of Catalyst Retention on Inventory Replacement," *Oil & Gas Journal,* December 26, 1983, pp. 141–145.

8. Tamborski, G. A., Magnabosco, L. M., Powell, J. W., and Yoo, J. S., "Catalyst Technology Improvements Make SO_x Emissions Control Affordable," presented at Katalistiks' 6th Annual FCC Symposium, Munich, Germany, May 22–23, 1985.

9. Thiel, P. G., Blazek, J. J., "Additive R," Grace Davison *Catalagram,* No. 71, 1985.

10. Engelhard Corporation, "Reduced Unit Cell Size Catalysts Offer Improved Octane for FCC Gasoline," *The Catalyst Report,* TI-762.

11. Engelhard Corporation, "Increasing Motor Octane by Catalytic Means, Part 2," *The Catalyst Report,* EC6100P.

12. Engelhard Corporation, "The Chemistry of FCC Coke Format," *The Catalyst Report,* Vol. 7, Issue 2.

13. Majon, R. J. and Spielman, J., "Increasing Gasoline Octane and Light Olefin Yields with ZSM-5," *The Catalyst Report,* Vol. 5, Issue 9.

14. Davison Div., W. R. Grace & Co., Grace Davison *Catalagram,* No. 72, 1985.

CHAPTER 4

Chemistry of FCC Reactions

A complex series of reactions (Table 4-1) take place when a large gas-oil molecule comes in contact with a 1200°F to 1400°F FCC catalyst. The distribution of products depends on many factors, including the nature and strength of the catalyst acid sites. Although most cracking is catalytic, certain thermal cracking reactions also occur. Factors such as nonideal contact between feedstock and catalyst in the riser and lack of prompt separation of cracked products in the reactor housing can contribute to the occurrence of thermal cracking. The purpose of this chapter is:

- to provide a general discussion of the chemistry of cracking (both thermal and catalytic).
- to highlight the role of the catalyst, and in particular, the influence of zeolites.
- to explain how cracking reactions affect the unit's heat balance.

Catalytic and thermal cracking proceed via different routes. A clear understanding of the different mechanisms involved is beneficial in areas such as: selecting the "right" catalyst for a given operation, troubleshooting unit operation properly, and developing a new catalyst formulation. Topics discussed in this chapter are:

- Thermal Cracking
- Catalytic Cracking
- Thermodynamic Aspects

4.1 THERMAL CRACKING

Before the advent of cracking catalysts, thermal cracking was the primary process available to petroleum refiners for conversion of

Table 4-1
Important Reactions Occurring in FCC

1. Cracking:

   Paraffins cracked to olefins
   and smaller paraffins $\qquad$ $C_{10}H_{22} \rightarrow C_3H_6 + C_7H_{16}$

   Olefins cracked to smaller olefins $\quad$ $C_8H_{16} \rightarrow C_5H_{10} + C_3H_6$

   Aromatic side-chain scission $\quad$ $ArC_{10}H_{21} \rightarrow ArC_5H_9 + C_5H_{12}$

   Naphthenes (cycloparaffins)
   cracked to olefins and smaller
   ring compounds $\qquad$ Cyclo-$C_{10}H_{20} \rightarrow C_6H_{12} + C_4H_8$

2. Isomerization:

   Normal olefins to iso-olefin $\quad$ 1-$C_4H_8 \rightarrow$ trans-2-C_4H_8

   Normal paraffins to isoparaffin $\quad$ n-$C_4H_{10} \rightarrow$ iso-C_4H_{10}

3. Hydrogen Transfer: $\qquad$ Naphthene + Olefin
   $\rightarrow$ Aromatic + Paraffin

   Cyclo-aromatization $\qquad$ $C_6H_{12} + 3C_5H_{10} \rightarrow C_6H_6 + 3C_5H_{12}$

   Olefins to paraffins and aromatic $\quad$ $4C_6H_{12} \rightarrow 3C_6H_{14} + C_6H_6$

4. Transalkylation/Alkyl-group
   Transfer $\qquad$ $C_6H_4(CH_3)_2 + C_6H_6 \rightarrow 2C_6H_5CH_3$

5. Cyclization of Olefins to
   Naphthenes $\qquad$ $C_7H_{14} \rightarrow CH_3$-cyclo-$C_6H_{11}$

6. Dehydrogenation $\qquad$ n-$C_8H_{18} \rightarrow C_8H_{16} + H_2$

7. Dealkylation $\qquad$ iso-C_3H_7-$C_6H_5 \rightarrow C_6H_6 + C_3H_6$

8. Condensation $\qquad$ Ar-CH=CH$_2$ + R$_1$CH=CHR$_2$
   $\rightarrow$ Ar—Ar + 2H

low-value feedstocks. Even today, refiners are employing thermal processes such as "delayed coking" and "visbreaking" for cracking of residual hydrocarbons.

Cracking of a hydrocarbon means the breaking of a carbon to carbon bond. Thermal cracking is a function of temperature and time. The reaction occurs when hydrocarbons in the absence of a catalyst are exposed to high temperatures in the range of 800°F to 1200°F.

The initial step in the chemistry of thermal cracking is the formation of *free radicals*. A free radical is an uncharged molecule which has an unpaired electron. Free radicals are very reactive and short lived. They are formed upon splitting the C—C bond. The rupturing produces two uncharged species which share a pair of electrons. Equation 4-1 shows formation of free radical when a paraffin molecule is thermally cracked:

$$
\begin{array}{ccccc}
H & H & & H & H \\
| & | & & | & | \\
R_1 - C - C - R_2 & \rightarrow & R_1 - C^{\cdot} + {}^{\cdot}C - R_2 \\
| & | & & | & | \\
H & H & & H & H
\end{array}
\qquad (4\text{-}1)
$$

Free-radicals are extremely reactive. They can undergo alpha and beta scission and/or polymerization. The beta-scission produces an olefin and a primary free radical (Equation 4-2) which has two fewer carbon atoms [1]:

$$
R - CH_2 - CH_2 - {}^{\cdot}C - H_2 \rightarrow R - {}^{\cdot}C - H_2 \quad + H_2C = CH_2 \quad (4\text{-}2)
$$

Beta-scission is cracking at two bonds away from the free radical. The newly formed primary free radical can further undergo beta-scission to yield ethylene.

Alpha-scission will yield a methyl radical, which can abstract a hydrogen atom from a neutral hydrocarbon molecule. The hydrogen abstraction produces methane and a secondary or tertiary free radical (Equation 4-3):

$$
H_3C^{\cdot} + R\text{-}CH_2\text{-}CH_2\text{-}CH_2\text{-}CH_2\text{-}CH_2\text{-}CH_2\text{-}CH_3
$$

$$
\rightarrow CH_4 + R\text{-}CH_2\text{-}CH_2\text{-}CH_2\text{-}CH_2\text{-}{}^{\cdot}CH\text{-}CH_2\text{-}CH_3 \qquad (4\text{-}3)
$$

Once again, this radical can undergo beta-scission. The products will be an alpha-olefin and a primary free radical (Equation 4-4):

$$
R\text{-}CH_2\text{-}CH_2\text{-}CH_2\text{-}CH_2\text{-}{}^{\cdot}CH\text{-}CH_2\text{-}CH_3 \rightarrow R\text{-}CH_2\text{-}CH_2\text{-}{}^{\cdot}CH_2
$$

$$
+ H_2C{=}CH\text{-}CH_2\text{-}CH_3 \qquad (4\text{-}4)
$$

Similar to the methyl radical, the R-$\cdot$CH$_2$ radical can also abstract a hydrogen atom from another paraffin to form a secondary free radical and a smaller paraffin (Equation 4-5):

$$R_1\text{-}^{\cdot}CH_2 + R\text{-}CH_2\text{-}CH_2\text{-}CH_2\text{-}CH_2\text{-}CH_2\text{-}CH_2\text{-}CH_3 \rightarrow R_1\text{-}CH_3$$

$$+ R\text{-}CH_2\text{-}CH_2\text{-}CH_2\text{-}CH_2\text{-}CH_2\text{-}^{\cdot}CH\text{-}CH_3 \qquad (4\text{-}5)$$

R-$\cdot$CH$_2$ is more stable than H$_3\cdot$C. Consequently, the hydrogen abstraction rate of R-$\cdot$CH$_2$ is lower than that of the methyl radical.

The above sequence of reactions will lead to formation of a final product that is rich in C$_1$ and C$_2$ as well as a fair amount of alpha olefins. Free radicals undergo little branching (isomerization). One of the drawbacks of thermal cracking is that a high percentage of the olefins that are formed during intermediate reactions polymerize and condense directly to coke. The product distribution from thermal cracking is different from catalytic cracking, as shown in Table 4-2. The shifts in distribution

Table 4-2
Comparison of Major Features of Thermal and Catalytic Cracking

Hydrocarbon Type	Thermal Cracking	Catalytic Cracking
n-Paraffins	C$_2$ is major product, with much C$_1$ and C$_3$, and C$_4$ to C$_{16}$ olefins; little branching	C$_3$ to C$_6$ is major product; few n-olefins above C$_4$; much branching
Olefins	Slow double-bond shifts and little skeletal isomerization; H-transfer is minor and nonselective for tertiary olefins; only small amounts of aromatics formed from aliphatics at 932°F	Rapid double-bond shifts, extensive skeletal isomerization, H-transfer is major and selective for tertiary olefins; large amounts of aromatics formed from aliphatics at 932°F
Naphthenes	Crack at slower rate than paraffins	If structural groups are equivalent, crack at about the same rate as paraffins
Alkylaromatics	Cracked within side chain	Cracking next to ring is prominent

Source: Venuto [2]

of products clearly confirm the fact that these two processes must proceed via different mechanisms.

4.2 CATALYTIC CRACKING

The catalytic reactions can be classified into two broad categories: primary cracking of the gas oil molecule and secondary rearrangement and recracking of cracked products.

Before discussing mechanisms of the reactions, it is appropriate to provide a historical review of FCC catalyst development and to examine its cracking properties. An in-depth discussion of FCC catalyst is presented in Chapter 3.

4.2.1 FCC Catalyst Development

The first commercial fluidized cracking catalyst was acid-treated natural clay. Later, synthetic silica-alumina materials containing 10 to 15 percent alumina replaced the natural clay catalysts. The synthetic silica-alumina catalysts were more stable and yielded superior products. In the mid-1950s, alumina-silica catalysts containing 25 percent alumina came into use because of their higher stability. These synthetic catalysts were amorphous, meaning that their structure consisted of a random array of silica and alumina tetrahedrally connected. Some minor improvements in yields and selectivity were achieved by switching to the catalysts such as magnesia-silica and alumina-zirconia-silica.

Impact of zeolites

The breakthrough in FCC catalysts was the use of X and Y zeolites in the catalyst in early 1960s. The addition of these zeolites substantially increased catalyst activity and selectivity. The product distribution from a zeolite-containing catalyst is different from an amorphous silica-alumina catalyst (Table 4-3). Zeolites are significantly more active (over 1,000 times) than the amorphous silica-alumina. The higher activity comes mainly from greater strength and organization of the active sites in the zeolites.

Zeolites are crystalline alumina-silicates having a regular pore structure. Their basic building blocks are silica and alumina tetrahedra. Each tetrahedron consists of silicon or aluminum atoms at the center of the tetrahedron with oxygen atoms at the corners. Because silicon and aluminum are in a +4 and +3 oxidation state respectively, there

Table 4-3
Comparison of Yield Structure for Fluid Catalytic Cracking
of Waxy Gas Oil over Commercial Equilibrium Zeolite
and Amorphous Catalysts

Yields, at 80 vol% Conversion	Amorphous, High Alumina	Zeolite, XZ-25	Change from Amorphous
Hydrogen, wt%	0.08	0.04	–0.04
C_1's + C_2's, wt%	3.8	2.1	–1.7
Propylene, vol%	16.1	11.8	–4.3
Propane, vol%	1.5	1.3	–.02
Total C_3's	17.6	13.1	–4.5
Butenes, vol%	12.2	7.8	–4.4
i-Butane, vol%	7.9	7.2	–0.7
n-Butane, vol%	0.7	0.4	–0.3
Total C_4's	20.8	15.4	–5.4
C_5-390 at 90% ASTM gasoline, vol%	55.5	62.0	+6.5
Light Fuel Oil, vol%	4.2	6.1	+1.9
Heavy Fuel Oil, vol%	15.8	13.9	–1.9
Coke, wt%	5.6	4.1	–1.5
Gasoline Octane No.	94	89.8	–4.2

Source: Venuto [2]

is a net charge of –1 which must be balanced by a *cation* to maintain electrical neutrality.

The activity and selectivity of zeolites largely depend on the type of cations that occupy the zeolite structure. FCC zeolites are synthesized in an alkaline environment such as sodium hydroxide. The soda Y zeolites have little or no stability. However, these alkali cations can be easily exchanged. The acidity of the zeolite's active site is enhanced upon ion exchanging of sodium with cations such as hydrogen or rare earth ions. The most widely used rare earth compounds are lantheium (La^{3+}) and cerium (Ce^{3+}).

The catalyst acid sites are both Bronsted and Lewis type. The catalyst can have either strong or weak Bronsted sites or strong or weak

Lewis sites. The catalyst acid properties depend on several parameters, including method of preparation, dehydration temperature, silica-to-alumina ratio, and the ratio of Bronsted to Lewis acid sites.

Bronsted's definition of an acid is a substance capable of donating a proton. Bronsted acid is the traditional hydrogen donor acid, such as hydrochloric and sulfuric acids. Lewis' definition of an acid is a substance that accepts a pair of electrons. Lewis acids may not have hydrogen in them but they are still acids. The classical example is aluminum chloride. Aluminum chloride in water will react with hydroxyl, causing a drop in solution pH.

4.2.3 Mechanism of Catalytic Cracking Reactions

When the feed contacts the regenerated catalyst, the first step is the vaporization of the feed by the catalyst. The next step is formation of positive-charged carbon atoms called *carbocations*. Carbocation is a generic term for a positive-charged carbon ion. Carbocation can further be subdivided into *carbenium* and *carbonium* ions.

A carbenium ion, $R\text{-}CH_2^+$, comes either from adding a positive charge to an olefin and/or from removing a hydrogen and two electrons from a paraffin molecule (Equations 4-6 and 4-7).

$$R - CH = CH - CH_2 - CH_2 - CH_3 + H^+ \text{ (a proton @ Bronsted site)}$$

$$\rightarrow R - C^+H - CH_2 - CH_2 - CH_2 - CH_3 \qquad (4\text{-}6)$$

$$R - CH_2 - CH_2 - CH_2 - CH_3 \text{ (removal of } H^- \text{ @ Lewis site)}$$

$$\rightarrow R - C^+H - CH_2 - CH_2 - CH_3 \qquad (4\text{-}7)$$

The Bronsted and Lewis acid sites on the catalyst are responsible for generating carbenium ions. The Bronsted site donates a proton to an olefin molecule and the Lewis site removes electrons from a paraffin molecule. In commercial units, olefins are either in the feed or are produced through thermal cracking reactions.

A carbonium ion, CH_5^+, is formed by adding a hydrogen ion (H^+) to a paraffin molecule (Equation 4-8). This is accomplished via direct

attack of a proton from the catalyst Bronsted site. The resulting molecule will have a positive charge with 5 bonds to it.

$$R — CH_2 — CH_2 — CH_2 — CH_3 + H^+ \text{ (proton attack)}$$

$$→ R — C^+H — CH_2 — CH_2 — CH_3 + H_2 \qquad (4\text{-}8)$$

The carbonium ion's charge is not stable and the catalyst acid sites are
probably not strong enough to form a lot of carbonium ions. Consequently, nearly all the cat cracking chemistry is carbenium ion chemistry.

The stability of carbocations depends on the nature of alkyl groups
attached to the positive charge. The relative stability of carbenium ions
is as follows [2]:

$$
\begin{array}{ccccccccc}
\text{Tertiary} & > & \text{Secondary} & > & \text{Primary} & > & \text{Ethyl} & > & \text{Methyl} \\
R — C — C^+ — C & & C — C^+ — C & & R — C — C^+ & & C — C^+ & & C^+ \\
\quad\; | & & & & & & & & \\
\quad\; C & & & & & & & &
\end{array}
$$

One of the benefits of catalytic cracking is that the primary and
secondary ions tend to rearrange to form a *tertiary* ion (a carbon with
three other carbon bonds attached). As will be discussed later, the
increased stability of tertiary ions accounts for the high degree of
branching associated with cat cracking.

Once formed in the initial step, carbenium ions can form a number
of different reactions. The nature and strength of the catalyst acid sites
will significantly influence the degree to which these reactions occur.
The three dominant reactions of carbenium ions are:

- The cracking of a carbon-carbon bond
- Isomerization
- Hydrogen transfer

Cracking reactions

The cracking, or beta-scission, is a key feature of ionic cracking.
Beta-scission is splitting of the C-C bond at two bonds away from
the positive-charge carbon atom. There is a preference for beta-scission
because the energy required to break this bond is lower than that
needed to break the adjacent C-C bonds. In addition, long-chain

hydrocarbons are more reactive than short-chain hydrocarbons; therefore, the rate of the cracking reactions decreases with decreasing chain length to the point that it is not possible to form stable carbenium ions.

The initial products of beta-scission are an olefin and a new carbenium ion (Equation 4-9). The newly formed carbenium ion will then continue a series of chain reactions. Small ions such as a four-carbon or five-carbon can then react with another big molecule and transfer the positive charge, and then the big molecule can crack. Cracking does not eliminate the positive charge; it stays there until two ions run into each other. The smaller ions are more stable and will not crack. They stay longer and finally transfer their charge into a big molecule.

$$R - C^+H - CH_2 - CH_2 - CH_2 - CH_3$$

$$\rightarrow CH_3 - CH = CH_2 \quad + C^+H_2 - CH_2 - CH_2R \qquad (4\text{-}9)$$

Because beta-scission is monomolecular, cracking is endothermic. Consequently, cracking rate is favored by high temperatures; cracking is not equilibrium limited.

Isomerization reactions

Isomerization reactions occur more frequently in catalytic cracking than in thermal cracking. As discussed earlier, thermal cracking is a free-radical mechanism. Breaking of a bond in both thermal and catalytic mechanisms is via beta-scission; however, in catalytic cracking, a number of carbocations tend to rearrange to form tertiary ions. Tertiary ions are more stable than secondary and primary ions; they shift around and crack to produce branched molecules (Equation 4-10). Free radicals do not do that; they yield normal or straight compounds.

$$CH_3 - CH_2 - C^+H - CH_2 - CH_2R \rightarrow CH_3 - \underset{\underset{H}{|}}{C^+} - \underset{\underset{CH_3}{|}}{CH} - CH_2R$$

$$\text{or}$$

$$C^+H_2 - \underset{\underset{CH_3}{|}}{CH} - CH_2 - CH_2R \qquad (4\text{-}10)$$

Some of the advantages of isomerization are as follows:

- Higher octane
- Higher-value chemical and oxygenate feedstocks
- Lower cloud point for diesel fuel

The isoparaffins in the gasoline boiling range have higher octane than normal paraffins. Compounds such as isobutylene and isoamylene will be extremely valuable as the feedstocks for the production of methyl tertiary butyl ether (MTBE) and tertiary amyl methyl ether (TAME). MTBE and TAME can be blended into the gasoline to reduce auto emissions. Finally, isoparaffins in the light cycle oil boiling range improve the cloud point.

Hydrogen transfer reactions

Hydrogen transfer, or more correctly hydride transfer, is a bimolecular reaction in which one reactant is an olefin. An example of hydrogen transfer is the reaction of two olefins. Both olefins would have to be adsorbed on the active sites and the sites would have to be close together for these reactions to take place. One of theses olefins becomes paraffin and the other becomes cyclo-olefins; so hydrogen is moved from one to another. Cyclo-olefin is now hydrogen transferred with another olefin to yield cyclodi-olefin. Cyclodi-olefin will then rearrange to form an aromatic, and aromatics are extremely stable. Therefore, hydrogen transfer of olefins converts them to paraffins and aromatics (Equation 4-11).

$$4\ C_nH_{2n} \rightarrow 3\ C_n\ H_{2n+2} + C_nH_{2n-6}$$

$$\text{olefins} \rightarrow \text{paraffins} + \text{aromatic} \tag{4-11}$$

Naphthenic compounds are also hydrogen donors and can react with olefins to produce paraffins and aromatics (Equation 4-12).

$$3\ C_nH_{2n} + C_mH_{2m} \rightarrow 3\ C_n\ H_{2n+2} + C_m\ H_{2m-6}$$

$$\text{olefins} + \text{naphthene} \rightarrow \text{paraffins} + \text{aromatic} \tag{4-12}$$

A rare-earth-exchanged zeolite increases indirectly hydrogen transfer reactions. In simple terms, the rare earth forms bridges between two to three acid sites in the catalyst framework. In doing so, the rare earth basically protects those acid sites being ejected from the framework. Because hydrogen transfer is promoted from adjacent acid sites, bridging these sites with rare earth promotes hydrogen transfer reactions.

Hydrogen transfer reactions usually increase FCC gasoline yield and its stability. It does so by reducing reactivity of the gasoline being produced. When there is hydrogen transfer, there are fewer olefins. Olefins are the reactive species in gasoline for secondary reactions; therefore, hydrogen transfer reactions reduce indirectly "overcracking" of the gasoline.

Some of the drawbacks of hydrogen transfer reactions are: lower gasoline octane, lower light olefins in the LPG, higher aromatics in the gasoline and LCO, as well as lower olefins in the front end of gasoline. The light (C_3's, C_4's, C_5's) olefins can further be processed in the alkylation and etherification units to yield excellent blendstocks for reformulated gasoline.

Cracking, isomerization, and hydrogen transfer reactions account for the majority of reactions occurring in cat cracking. There are other associated reactions that do indeed play an important role in unit operation. Two prominent reactions are dehydrogenation and coking. Under ideal conditions, i.e., a "clean" feedstock and catalyst with no metals, cat cracking does not yield any appreciable amounts of molecular hydrogen. Therefore, dehydrogenation reactions will only proceed if the catalyst is contaminated with metals such as nickel and vanadium.

Cat cracking of gas oil molecules yields a residue called coke. Chemistry of coke formation is complex and not very well understood. Similar to hydrogen transfer reactions, catalytic coke is a "bimolecular" reaction product and proceeds via carbenium ions or free radicals. In theory, coke yield should increase as hydrogen transfer rate is increased. It is postulated [4] that reactions producing unsaturates and multi-ring aromatics are the principal coke-forming compounds. Unsaturates such as olefins, diolefins, and multi-ring polycyclic olefins are very reactive and can polymerize to form coke.

For a given catalyst and feedstock, catalytic coke yield is a direct function of conversion. However, there exists an optimum riser temperature in which the coke yield is minimum. For a typical cat cracker, this temperature is about 950°F. Let's look at two extreme riser temperatures of 850°F and 1050°F. At 850°F, a lot of coke is formed mainly because the carbenium ions do not desorb at this low temperature. At 1050°F, a large amount of coke is formed largely due to olefin polymerization.

4.3 THERMODYNAMIC ASPECTS

As stated earlier, catalytic cracking involves a series of simultaneous reactions. Some of these reactions are endothermic and some are exothermic. Each reaction has a heat of reaction associated with it (Table 4-4). The overall heat of reaction refers to the net or combined heat of reaction of all the reactions. Although there are a number of exothermic reactions, the net reaction is still endothermic.

The regenerated catalyst supplies enough energy to heat the feed to the riser outlet temperature, to heat the combustion air to the flue gas temperature, to provide the endothermic heat of reaction, and to compensate for any heat losses to atmosphere. The source of this energy is the burning of coke produced from the reaction. It is apparent that the type and magnitude of these reactions have an impact on the heat balance of the unit. For example, a catalyst with less hydrogen transfer characteristics will cause the net heat of reaction to be more endothermic. Consequently, this will require a higher catalyst circulation and possibly a higher coke yield to maintain the heat balance.

SUMMARY

Although cat cracking reactions are predominantly catalytic, some nonselective thermal cracking reactions do take place. The difference in the distribution of products clearly confirms that these two processes proceed via different chemistry.

The introduction of zeolites into the FCC catalyst in the early 1960s drastically improved the performance of the cat cracker reaction products. The catalyst acid sites, their nature and strength, have a major influence on the reaction chemistry.

Catalytic cracking proceeds mainly via carbenium ion intermediates. The three dominant reactions are cracking, isomerization, and hydrogen transfer. Finally, the type and degree of reactions occurring in FCC will influence the unit heat balance.

Table 4-4

Some Thermodynamic Data for Idealized Reactions of Importance in Catalytic Cracking

Reaction Class	Specific Reaction	Log K_E (equilibrium constant)			Heat of Reaction BTU/mole
		850°F	950°F	980°F	950°F
Cracking	n-$C_{10}H_{22} \rightarrow$ n-C_7H_{16} + C_3H_6	2.04	2.46	—	32,050
	1-$C_8H_{16} \rightarrow$ 2C_4H_8	1.68	2.10	2.23	33,663
Hydrogen transfer	4$C_6H_{12} \rightarrow$ 3C_6H_{14} + C_6H_6	12.44	11.09	—	109,681
	cyclo-C_6H_{12} + 3 1-$C_5H_{10} \rightarrow$ 3n-C_5H_{12} + C_6H_6	11.22	10.35	—	73,249
Isomerization	1-$C_4H_8 \rightarrow$ trans-2-C_4H_8	0.32	0.25	0.09	-4,874
	n-$C_4H_{10} \rightarrow$ iso-C_4H_{10}	-0.20	-0.23	-0.36	-3,420
	o-$C_6H_4(CH_3)_2 \rightarrow$ m-$C_6H_4(CH_3)_2$	0.33	0.30	—	-1,310
	cyclo-$C_6H_{12} \rightarrow$ CH_3-cyclo-C_5H_9	1.00	1.09	1.10	6,264
Transalkylation	C_6H_6 + m-$C_6H_4(CH_3)_2 \rightarrow$ 2$C_6H_5CH_3$	0.65	0.65	0.65	-221
Cyclization	1-$C_7H_{14} \rightarrow$ CH_3-cyclo-C_6H_{11}	2.11	1.54	—	-37,980
Dealkylation	iso-C_3H_7-$C_6H_5 \rightarrow$ C_6H_6 + C_3H_6	0.41	0.88	1.05	40,602
Dehydrogenation	n-$C_6H_{14} \rightarrow$ 1-C_6H_{12} + H_2	-2.21	-1.52	—	56,008
Polymerization	3$C_2H_4 \rightarrow$ 1-C_6H_{12}	—	—	-1.2	—
Paraffin Alkylation	1-C_4H_8 + iso-$C_4H_{10} \rightarrow$ iso-C_8H_{18}	—	—	03.30	—

Source: Venuto [2]

REFERENCES*

1. Gates, B. C., Katzer, J. R., and Schuit, G. G., *Chemistry of Catalytic Processes.* New York: McGraw-Hill, 1979.
2. Venuto, P. B. and Habib, E. T., *Fluid Catalytic Cracking with Zeolite Catalysts.* New York: Marcel Dekker, Inc., 1979.
3. Broekhoven, E. V. and Wijngaards, H., "Investigation of the Acid Site Distribution of FCC Catalysts with Ortho-xylene as a Model Compound," 1988 Akzo Chemicals FCC Symposium, Amsterdam, The Netherlands.
4. Koermer, G. and Deeba, M., "The Chemistry of FCC Coke Formation," Engelhard Corporation, *The Catalyst Report,* Vol. 7, Issue 2, 1991.

* The author also expresses appreciation to Messrs. Terry Reid of Akzo Nobel and Tom Habib of Davison Div., W. R. Grace & Co., for their many helpful comments.

CHAPTER 5

Unit Monitoring and Control

The only proper way to monitor and evaluate performance of a cat cracker is by conducting periodic material and heat balance surveys on the unit. By carrying out these tests frequently, one can collect, trend, and evaluate the unit operating data systematically. Additionally, meaningful technical service to optimize FCC unit operation should be based on regular test run results.

Understanding the operation of a commercial cat cracker also requires an in-depth knowledge of the unit's heat balance requirements. Any changes to feedstock quality, operating conditions, catalyst, and mechanical configuration will impact the heat balance of the unit. Heat balance is an important tool in predicting and evaluating the changes that will affect the quantity and the quality of FCC products.

Finally, before the unit can produce one barrel of product, it must circulate catalyst smoothly. Therefore, one must be familiar with the dynamics of pressure balance and key process controls.

The main topics discussed in this chapter are:

- Material Balance
- Heat Balance
- Pressure Balance
- Process Control Instrumentation

In the material and heat balance sections, the discussions include two methods for performing test runs, some practical steps for carrying out a successful test run, a step-by-step method for performing a material and heat balance, and an actual case study. In the pressure balance, and process control sections, the significance of the pressure balance in "debottlenecking" the unit's constraints is discussed, and finally, fundamentals of both "basic" and "advanced" process controls are presented.

5.1 MATERIAL BALANCE

A complete set of data collections should be carried out weekly. This will permit distinction between the effects of feedstock, catalyst, and operating conditions on unit performance. An accurate assessment of a cat cracker operation requires reliable plant data. A reasonable weight balance should have at least a 98%–102% closure.

In any weight balance exercise, the first step is to identify the input and output streams. This is usually done by drawing an envelope(s) around the input and output streams. Two examples of such envelopes are shown in Figure 5-1.

One of the main objectives of performing a material balance is to determine directly or indirectly the composition of products leaving the reactor. The reactor effluent vapor entering the main fractionator contains hydrocarbons, steam, and inert gases. By weight, the hydrocarbons in the reactor overhead stream are equal to the fresh feed minus the portion of the feed that has been converted to coke. This is true assuming that no recycle is processed to the reactor. If there is a recycle stream, its amount should be included with the reactor effluent stream.

The sources of steam in the reactor vapor are lift, dispersion, dome, stripping, and other miscellaneous steam to the reactor. Depending on the reactor pressure, approximately 25% to 50% of stripping steam is entrained with the spent catalyst flowing to the regenerator.

Inert gases such as nitrogen and carbon dioxide are entrained with the regenerated catalyst. The quantity of these inert gases is related directly to catalyst circulation rate. These gases flow through the gas plant and leave the unit with the off-gas from the sponge oil absorber column.

FCC products are commonly reported, on an inert-free basis, as the volume and weight fractions of the fresh feed. Additionally, gasoline, light cycle oil (LCO), and conversion are reported on a fixed basis such as true boiling point (TBP) cut point. The standard cut points are 430°F TBP end point for gasoline and 650°F TBP end point for LCO. Other popular cut points are 430°F ASTM D-86 end point for gasoline and 650°F or 670°F ASTM D-86 end point for LCO.

The "classic" definition of conversion is the volume or weight percent of feedstock converted to gasoline and other products lighter than gasoline. This includes coke yield. Conversion is calculated by:

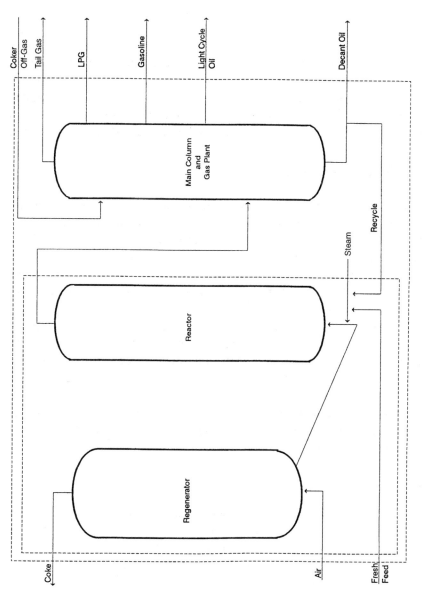

Figure 5-1. FCC unit input/output streams.

$$\text{Conversion \%} = \frac{\text{Feed} - (\text{light cycle oil} + \text{heavy cycle oil} + \text{decanted oil})}{\text{Feed}} \times 100$$

Depending on seasonal demands, the D-86 end point of an FCC gasoline can range from 380°F to 450°F. Any undercutting of gasoline below the standard cut point goes to the LCO product; therefore, it is necessary to distinguish between the *apparent* and *true* conversion. The apparent conversion is calculated before the gasoline cut point adjustment is made, and the true conversion is calculated after the cut point adjustment.

5.1.1 Testing Methods

The material balance around a cat cracker can be carried out using two different sampling techniques. The first method, practiced by the majority of refineries, is to conduct a unit test run in which the flow rates and laboratory analyses of the feed and product streams are obtained. Because, in many FCC units, streams from other units are also fed to the FCC recovery section, their flow rates and composition must also be known.

The second method involves direct sampling of the reactor effluent as shown in Figure 5-2. In this technique, a sample of total reactor effluent is withdrawn and collected in an aluminized polyester bag for separation and further analysis.

There are several advantages and disadvantages to reactor effluent sampling, as shown below. For both sampling methods, the amount of air to the regenerator and the flue gas composition must be known to calculate coke yield.

Advantages

- Allows data gathering on different sets of conditions without the whole recovery side to equilibrate.
- Eliminates concern about rate and compositions of extraneous streams entering the gas plant because they are not included in the overall balance.
- Eliminates concern about correcting for cut points because the effluent sample is cut at the desired TBP cut point.
- Eliminates concern about obtaining a 100% weight balance.

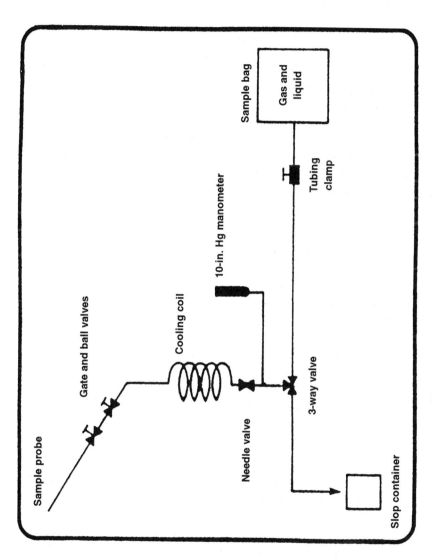

Figure 5-2. Reaction mix sampling [2].

Disadvantages

- Possible leaks during sampling.
- Possible inaccurate measurement of volume of gas and weight of liquid.
- Requires qualified individuals to perform the test.
- Requires separate lab to perform analyses.

5.1.2 Recommended Procedures
for Conducting a Test Run

Conducting a successful cat cracker test run requires a clear definition of objectives, careful planning, and proper interpretation of the results. The following steps can be used as a guide to ensure a smooth and successful test run.

Prior to the test run

1. A memo should be issued to the involved departments, such as operations, laboratory, and oil movement, communicating the purpose, duration, and scope of the test run. Additionally, a list of samples and the required analysis (Table 5-1) should be furnished.
2. The composition of the crude oil and FCC feedstock should remain relatively constant during the test run.
3. The key flow meters should be zeroed and calibrated.
4. It should be verified that the sample taps are not plugged.
5. The "sample bombs" used to collect gas and LPG products should be purged.

Data collection

1. The test run duration should be specified—usually 8 to 12 hours. It should be documented which constraints—blower, wet gas compressor, etc.—the unit is operating against.
2. The "sample taps" must be bled adequately before samples are collected. Because of the importance of obtaining a reliable flue gas analysis, it is suggested that an extra sample of flue gas be collected. Additionally, the laboratory should be requested to retain the unused samples until all analyses are verified.

Table 5-1
Typical Laboratory Analysis of FCC Streams

	Tests						
	°API	D-86	D-1160	Sulfur	Viscosity	Metals	GC
Gas Oil Feedstock	✔		✔	✔	✔	✔	
Slurry Recycle	✔		✔	✔			
Decanted Oil Product	✔		✔	✔	✔		
LCO Product	✔	✔		✔			
Gasoline Product	✔	✔		✔			✔
LPG C$_3$'s and C$_4$'s				✔			✔
Fuel Gas							✔
E-Cat							

3. Pertinent operating data must be collected. A form similar to the one shown in Table 5-2 can be used to gather the data.

Mass balance calculations

1. The orifice plate meter factors should be adjusted for actual operating parameters. For liquid streams, the flow meters should be adjusted for °API gravity, temperature, and viscosity. For gas streams, the flow rate should be adjusted for the operating temperature, pressure, and molecular weight.
2. It is necessary to ensure that the chromatograph analyses of each stream are normalized to 100%.
3. By using air rate and flue gas composition, the coke yield should be calculated.

Table 5-2
Operating Data

Feed and Product Rates
Fresh Feed rate	50,000 bpd
Coker off gas	3,000,000 scfd
FCC Tail gas	16,000,000 scfd
LPG	11,565 bpd
Gasoline	30,000 bpd
LCO	10,000 bpd
DO	3,000 bpd

Other Pertinent Flow Rates
Dispersion Steam	9,000 lbs/hr
Reactor Stripping Steam	13,000 lbs/hr
Reactor Dome Steam	1,200 lbs/hr
Air to Regenerator	90,000 scf/min

Temperature, °F
Riser Inlet	594
Riser Outlet	972
Blower Discharge	374
Regen. Dense Phase	1309
Regen. Flue Gas	1330
Ambient	80

Pressure, psig
Blower Discharge	43
Regen. Dome	34
Reactor Dome	33
Regenerated Catalyst Slide Valve, ΔP	5.8
Spent Catalyst Slide Valve, ΔP	6.0

Flue Gas Analysis, Mol%
O_2	1.5
CO_2	15.4
CO	0.0
SO_2	500 ppm $\rightarrow$ 0.05 mol%
$N_2 + Ar$	83.05

Miscellaneous Data
Relative Humidity	80%
Fresh Catalyst Makeup	4 tons/day
E-cat MAT	68%

4. The flow rate of each stream should be converted to the unit of weight.
5. The quantity of inert gases and extraneous streams should be subtracted from the FCC gas plant products.
6. After calculating the overall mass balance error, the feed/products should be mass-balanced either by redistributing the error in proportion to their rates or adjusting a known inaccurate meter(s).
7. The actual gasoline and LCO cut points should then be converted to standard cut points.
8. Using the feed characterization correlation discussed in Chapter 2, composition of fresh feedstock can be determined.

5.1.3 Analysis of Results

1. The performance of the cat cracker in regard to the yields and quality of the desired products should be assessed.
2. The results of this test run should be compared with the previous test runs; any significant changes in the yields and/or operating parameters should be highlighted.
3. The final step is to perform simple economics of the unit operation and make recommendations that improve unit operation short and long term.

The following case study demonstrates a step-by-step approach to performing a comprehensive material and heat balance.

5.1.4 Case Study

A test run is conducted to evaluate performance of a 50,000 bpd FCC unit. The feed to the unit is gas oil from the vacuum unit. No recycle stream is processed; however, the off-gas from the delayed coker is sent to the cat cracker recovery section. Products from the unit are fuel gas, LPG, gasoline, LCO, and decanted oil (DO). Tables 5-2 and 5-3 contain stream flow rates, operating data, and laboratory analyses. It is assumed that the meter factors have been adjusted for actual operating conditions.

The sequence of steps to perform the mass balance is as follows:

1. Identification of the input and output streams used in the overall mass balance equation.

Table 5-3
Feed and Products Inspections

	Feed	Gasoline	LCO	Decanted Oil
°API Gravity	25.2	58.5	21.5	2.4
Sulfur, Wt%	0.5			
Aniline Point, °F	208			
RI @ 67°C	1.4854			
Viscosity, SSU				
@ 150°F	109			
@ 210°F	54			
Distillation, °F	D-1160	D-86	D-86	D-1160
Vol%				
10	682	125	477	646
30	766	160	514	687
50	835	213	547	720
70	901	285	576	771
90	1001	369	627	846
EP	1060	433	666	1055

Mole% Composition of FCC Gas Plant Streams

Component	FCC Tail Gas	LPG	FCC Gasoline	Coker Off-Gas
H_2	15.5			8.0
CH_4	35.8			47.2
C_2	17.1			14.9
$C_2^=$	11.0			2.5
C_3	1.6	17.9		8.4
$C_3^=$	4.7	31.3		4.4
IC_4	0.7	16.1	0.4	0.9
NC_4	0.2	10.9	2.0	3.2
C_4	1.3	23.8	4.4	3.4
C_5+	1.0		93.2	4.9
H_2S	2.1			2.0
N_2	7.2			
CO_2	1.8			0.2
Total	**100.0**	**100.0**	**100.0**	**100.0**
Sp. Gravity	0.78	0.55		0.96

2. Calculation of the coke yield.
3. Conversion of the appropriate flow rates to the unit of mass, e.g., lbs/hr.
4. Normalization of the data to obtain a 100% weight balance.
5. Determination of the component yields.
6. Adjustment of the gasoline, LCO, and decanted oil yields to standard cut points.

Input and output streams in the overall mass balance

As shown in envelope I of Figure 5-1, the input hydrocarbon streams are fresh feed and coker off-gas. The output streams are FCC tail gas (minus inerts), LPG, gasoline, LCO, DO, and coke.

Coke yield calculations

As discussed in Chapter 1, a portion of the feed is converted to coke in the reactor. This coke is carried into the regenerator with the spent catalyst. The combustion of the coke produces H_2O, CO, CO_2, SO_2 and traces of NO_x. To determine coke yield, the amount of dry air to the regenerator and analysis of flue gas are needed. It is essential to have an accurate analysis of the flue gas. The hydrogen content of coke relates to the amount of hydrocarbon vapors carried over with the spent catalyst into the regenerator, and is an indication of the reactor-stripper performance. Example 5-1 shows a step-by-step calculation of the coke yield.

Example 5-1
Determination of the Unit's Coke Yield

Given: Wet air = 90,000 SCFM, Relative Humidity = 80%, Ambient Temperature = 80°F

Figure 5-3 can be used to obtain percent dry air as the function of ambient temperature and relative humidity. For this example, the percentage of dry air is 97.1% or:

$$\bullet \; \text{Dry Air} = 0.971 \times \frac{90,000 \text{ SCF}}{\text{Min}} \times \frac{1 \text{ mole}}{379 \text{ SCF}} \times \frac{60 \text{ Min}}{1 \text{ hr}} = 13,835 \text{ moles/hr}$$

Flue gas rate (dry basis) is calculated from dry air rate using nitrogen and argon as tic elements.

$$\bullet \ \text{Flue gas rate (dry basis)} = \frac{(13{,}835 \ \text{moles}/\text{hr} \times 0.7901)}{0.8305} = 13{,}162 \ \text{moles}/\text{hr}$$

0.7901 and 0.8305 are concentrations of (nitrogen + argon) in atmospheric dry air and flue gas, respectively.

The flow rates of each component in the flue gas stream are:

- O_2 out = 0.015 × 13,162 moles/hr = 197.4 moles/hr
- CO_2 out = 0.154 × 13,162 moles/hr = 2,026.9 moles/hr
- SO_2 out = 0.005 × 13,162 moles/hr = 6.6 moles/hr
- $(N_2 + Ar)$ out = 0.8305 × 13,162 moles/hr = 10,913 moles/hr

An oxygen balance can be used to calculate water formed by the combustion of coke:

- O_2 out = 197.4 + 2,026.9 + 6.6 = 2,231 moles/hr
- O_2 in = 0.2095 × 13,835 moles/hr = 2,894.4 moles/hr
- O_2 used during combustion = 2,898.4 – 2231.0 = 667.4 moles/hr

Since for each mole of O_2, two moles of water are formed, the amount of water is:

- H_2O formed = 667.4 × 2 = 1,334.8 moles/hr

Components of coke are carbon, hydrogen, and sulfur. Their rates are calculated as follows:

- Carbon = 2,026.9 moles/hr × 12 lbs/mole = 24,323 lbs/hr
- Hydrogen = 1,334.8 moles/hr × 2.02 lbs/mole = 2,696 lbs/hr
- Sulfur = 6.6 moles/hr × 32.1 lbs/moles = 212 lbs/hr
- Coke = 24,323 + 2,696 + 212 = 27,231 lbs/hr

- H_2 content of coke, wt% = $\dfrac{2{,}696 \ \text{lbs}/\text{hr}}{27{,}231 \ \text{lbs}/\text{hr}} \times 100 = 9.9$

Conversion to unit of weight, lbs/hr

The next step is to convert flow rate of each stream in the overall mass balance equation to the unit of weight, e.g., lbs/hr. Example 5-2 shows these conversions for the gas and liquid streams.

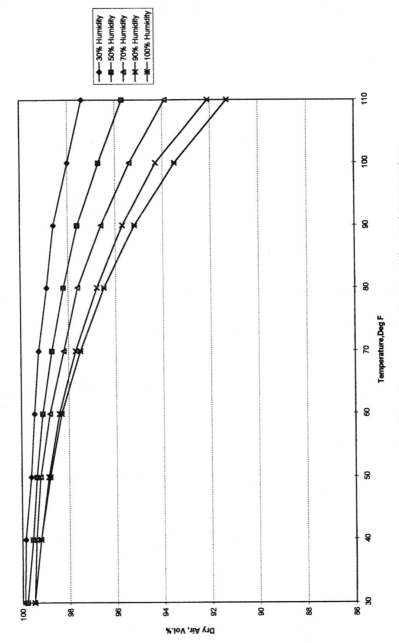

Figure 5-3. Dry air versus relative humidity and temperature.

Example 5-2
Conversion of Input and Output Streams to the Unit of Weight (lbs/hr)

- Fresh Feed $= \dfrac{50,000 \text{ bbl}}{\text{day}} \times \dfrac{1 \text{ day}}{24 \text{ hr}} \times \dfrac{141.5}{(131.5 + 25.2)} \times \dfrac{8.33 \text{ lbs}}{\text{gal}} \times \dfrac{42 \text{ gal}}{\text{bbl}}$

  $= 658,814 \text{ lbs/hr}$

- Coker gas $= \dfrac{3,000,000 \text{ SCF}}{\text{day}} \times \dfrac{1 \text{ day}}{24 \text{ hr}} \times \dfrac{1 \text{ mole}}{379.5 \text{ SCF}} \times \dfrac{27.8 \text{ lbs}}{1 \text{ mole}} = 9,157 \text{ lbs/hr}$

- FCC tail gas $= \dfrac{16,000,000 \text{ SCF}}{\text{day}} \times \dfrac{1 \text{ day}}{24 \text{ hr}} \times \dfrac{1 \text{ mole}}{379.5 \text{ SCF}} \times \dfrac{22.6 \text{ lbs}}{1 \text{ mole}}$

  $= 39,701 \text{ lbs/hr}$

The amount of inerts in the FCC tail gas is:

- $N_2 = \dfrac{16,000,000 \text{ SCF}}{\text{day}} \times \dfrac{1 \text{ day}}{24 \text{ hr}} \times 0.072 \times \dfrac{1 \text{ mole}}{379.5 \text{ SCF}} \times \dfrac{28 \text{ lbs}}{1 \text{ mole}} = 3,542 \text{ lbs/hr}$

- $CO_2 = \dfrac{16,000,000 \text{ SCF}}{\text{day}} \times 0.021 \times \dfrac{1 \text{ day}}{24 \text{ hr}} \times \dfrac{1 \text{ mole}}{379.5 \text{ SCF}} \times \dfrac{44 \text{ lbs}}{1 \text{ mole}} = 1,623 \text{ lbs/hr}$

- Inert-free FCC tail gas $= 39,701 - (3,542 + 1,623) = 34,537 \text{ lbs/hr}$

- LPG $= \dfrac{11,565 \text{ bbl}}{\text{day}} \times \dfrac{1 \text{ day}}{24 \text{ hr}} \times \dfrac{141.5}{131.5 + 123.5} \times \dfrac{8.34 \text{ lbs}}{\text{gal}} \times \dfrac{42 \text{ gal}}{\text{bbl}} = 93,662 \text{ lbs/hr}$

- Gasoline $= \dfrac{30,000 \text{ bbl}}{\text{day}} \times \dfrac{1 \text{ day}}{24 \text{ hr}} \times \dfrac{141.5}{131.5 + 58.5} \times \dfrac{8.34 \text{ lbs}}{\text{gal}} \times \dfrac{42 \text{ gal}}{\text{bbl}}$

  $= 326,083 \text{ lbs/hr}$

- LCO $= \dfrac{10,000 \text{ bbl}}{\text{day}} \times \dfrac{1 \text{ day}}{24 \text{ hr}} \times \dfrac{141.5}{131.5 + 21.5} \times \dfrac{8.34 \text{ lbs}}{\text{gal}} \times \dfrac{42 \text{ gal}}{\text{bbl}} = 134,980 \text{ lbs/hr}$

$$\bullet \; DO = \frac{3{,}000 \; bbl}{day} \times \frac{1 \; day}{24 \; hr} \times \frac{141.5}{131.5 + 2.4} \times \frac{8.34 \; lbs}{gal} \times \frac{42 \; gal}{bbl} = 46{,}270 \; lbs/hr$$

Normalization of the data

Because a preliminary weight balance seldom has a 100% closure, it is necessary to normalize the yield to obtain a 100% weight balance. Example 5-3 shows the preliminary overall weight balance.

Example 5-3
Preliminary Overall Weight Balance

Input = Fresh Feed + Coker Off-Gas
Output = FCC tail gas + LPG + Gasoline + LCO + DO + Coke

- Input = 658,814 + 9,182 = 667,996 lbs/hr

- Output = 34,617 + 93,656 + 326,124 + 134,973 + 46,270 + 27,231
    = 662,871 lbs/hr

- Difference = 667,996 − 662,871 = 5,125 lbs/hr

Error in mass balance = 0.8 wt%

The products are adjusted upward in proportion to their rates to obtain a 100% weight balance. The normalized rates:

- Tail gas = 34,883 lbs/hr
- LPG   = 94,460 lbs/hr = 11,658 bpd
- Gasoline = 328,766 lbs/hr = 30,230 bpd
- LCO   = 136,054 lbs/hr = 10,077 bpd
- DO   = 46,626 lbs/hr = 3,023 bpd
- Coke   = 27,440 lbs/hr

Component yield

The next task is to determine the reactor yield by performing a component balance. The amount of C_5+ in the gasoline boiling range is calculated by subtracting the C_4 and lighter components from the total gas plant products. Example 5-4 shows step-by-step calculation of the component yields. The summary of the results, normalized but unadjusted for the cut points, is shown in Table 5-4.

Example 5-4
Calculation of Individual Components

- $H_2 = \dfrac{0.155 \times 16 \text{ MMSCFD} \times 2.02}{379.5 \times 24} - \dfrac{0.08 \times 3 \text{ MMSCFD} \times 2.02}{379.5 \times 24} = 497 \text{ lbs/hr}$

- $CH_4 = \dfrac{0.358 \times 16 \text{ MMSCFD} \times 16}{379.5 \times 24} - \dfrac{0.472 \times 3.0 \text{ MMSCFD} \times 16}{379.5 \times 24} = 7,585 \text{ lbs/hr}$

- $C_2 = \dfrac{0.171 \times 16 \text{ MMSCFD} \times 30}{379.5 \times 24} - \dfrac{0.149 \times 3 \text{ MMSCFD} \times 30}{379.5 \times 24} = 7,549 \text{ lbs/hr}$

- $C_2^= = \dfrac{0.11 \times 16 \text{ MMSCFD} \times 28}{379.5 \times 24} - \dfrac{0.025 \times 3 \text{ MMSCFD} \times 28}{379.5 \times 24} = 5,187 \text{ lbs/hr}$

- $C_3 = \dfrac{0.016 \times 16 \text{ MMSCFD} \times 44}{379.5 \times 24} + \dfrac{0.179 \times 11,658 \text{ BPD} \times 175.3}{24}$

    $- \dfrac{0.084 \times 3 \text{ MMSCFD} \times 44}{379.5 \times 24} = 15,262 \text{ lbs/hr}$

- $C_3^= = \dfrac{0.047 \times 16 \text{ MMSCFD} \times 42}{379.5 \times 24} + \dfrac{0.313 \times 11,658 \text{ BPD} \times 181.8}{24}$

    $- \dfrac{0.044 \times 3 \text{ MMSCFD} \times 42}{379.5 \times 24} = 30,504 \text{ lbs/hr}$

- $NC_4 = \dfrac{0.002 \times 16 \text{ MMSCFD} \times 58}{379.5 \times 24} + \dfrac{0.109 \times 11,658 \text{ BPD} \times 204.6}{24}$

    $+ \dfrac{0.02 \times 30,230 \times 204.6}{24} - \dfrac{0.032 \times 3 \text{ MMSCFD} \times 58}{379.5 \times 24} = 15,579 \text{ lbs/hr}$

- $IC_4 = \dfrac{0.007 \times 16 \text{ MMSCFD} \times 58}{379.5 \times 24} + \dfrac{0.161 \times 11658 \text{ BPD} \times 197.2}{24}$

    $+ \dfrac{0.004 \times 30,230 \text{ BPD} \times 197.2}{24} - \dfrac{0.009 \times 3 \text{ MMSCFD} \times 58}{379.5 \times 24} = 16,958 \text{ lbs/hr}$

- $C_4^= = \dfrac{0.013 \times 16\ \text{MMSCFD} \times 56}{379.5 \times 24} + \dfrac{0.238 \times 11{,}658\ \text{BPD} \times 213.4}{24}$

$\qquad + \dfrac{0.044 \times 30{,}230\ \text{BPD} \times 213.4}{24} - \dfrac{0.034 \times 3\ \text{MMSCFD} \times 56}{379.5 \times 24} = 37{,}150\ \text{lbs/hr}$

Table 5-4
Normalized FCC Weight Balance Summary
with Coker Gas Subtracted

Stream	bpd	lbs/hr	Vol% of Feed	Wt% of Feed
Fresh Feed	50,000	658,814	100.00	100.00
Products				
H_2		497		0.07
C_1		7,585		1.15
C_2		7,549		1.15
$C_2^=$		5,187		0.79
Total C_2 and lighter		20,818		3.16
H_2S		1,032		0.16
C_3	2,090	15,262	4.18	2.32
$C_3^=$	4,027	30,504	8.05	4.63
IC_4	2,064	16,958	4.13	2.57
NC_4	1,827	15,579	3.65	2.36
$C_4^=$	4,178	37,150	8.36	5.64
Gasoline (C_5+)	28,650	311,437	57.30	47.27
LCO	10,077	136,008	20.15	20.64
DO	3,023	46,626	6.05	7.08
Coke		27,440		4.17
Total	55,936	658,814	111.87	100.00
Apparent Conversion			73.8	72.28
Inerts		5,143		

Adjustment of gasoline and LCO cut points

As discussed earlier in this chapter, gasoline and LCO yields are generally corrected to a constant boiling range basis. The most commonly used bases are 430°F TBP gasoline and 650°F TBP LCO end points. Since TBP distillations are not routinely performed, they often are estimated from the D-86 distillation data. The adjustment to the cut points involves the following:

- Adding to the raw LCO all the 430°F⁺ in the raw gasoline and subtracting the 430°F⁻.
- Adding to the raw LCO all the 650°F⁻ in the raw decanted oil and subtracting the 650°F⁻.
- Adding to the raw gasoline all the 430°F⁻ in the raw LCO and subtracting the 430°F⁺.
- Adding to the raw decanted oil all the 650°F⁺ in the raw LCO and subtracting the 650°F⁻.

Table 5-5 illustrates steps used to convert ASTM data to TBP. Extrapolation of the TBP data indicates the following:

- The 430°F⁺ content of the FCCU gasoline is 3 vol%, or 859 bpd.
- The gasoline (430°F⁻) content of LCO is 8 vol%, or 806 bpd.
- The 650°F⁺ content of LCO is 12 vol%, or 1,209 bpd.
- The LCO (650°F⁻) content of the decanted oil is 17 vol%, or 514 bpd.

Therefore, the adjusted rates are as follows:

Gasoline (C_5+ to 430°F TBP end point) = 28,650 − 859 + 806

= 28,597 bpd

LCO (430°F to 650°F TBP end point) = 10,077 + 514 − 1209 − 806

+ 859 = 9,435 bpd

DO (650°F⁺) = 3,023 + 1,209 − 514 = 3,718 bpd

Table 5-6 shows the normalized FCC weight balance with the adjusted cut points.

Table 5-5
Conversion of ASTM Distillation to TBP
Distillation for Gasoline, LCO and Decanted Oil

Gasoline TBP

(From Appendix 9, TBP 50% point = 213°F)

Given D-86	From Appendix 10
50% - 30% = 53°F	30% TBP = 140°F
30% - 10% = 35°F	10% TBP = 77°F
10% - IBP = 25°F	IBP TBP = 26°F
70% - 50% = 72°F	70% TBP = 297°F
90% - 70% = 84°F	90% TBP = 383°F
EP - 90% = 64°F	EP TBP = 501°F

LCO TBP

(From Appendix 9: TBP 50% point = 561°F)

Given D-86	From Appendix 10
50% - 30% = 33°F	30% TBP = 511°F
30% - 10% = 41°F	10% TBP = 441°F
10% - IBP = 73°F	IBP TBP = 343°F
70% - 50% = 29°F	70% TBP = 601°F
90% - 70% = 51°F	90% TBP = 660°F
EP - 90% = 39°F	EP TBP = 712°F

Decanted Oil TBP

(From Appendix 9: TBP 50% point = 744°F)

Given D-86	From Appendix 10
50% - 30% = 33°F	30% TBP = 694°F
30% - 10% = 41°F	10% TBP = 624°F
10% - IBP = 236°F	IBP TBP = 425°F
70% - 50% = 51°F	70% TBP = 807°F
90% - 70% = 75°F	90% TBP = 886°F

Table 5-6
Normalized and Adjusted FCC Weight Balance Summary

Stream	bpd	lbs/hr	Vol% of Feed	Wt% of Feed
Fresh Feed	50,000	658,814	100.00	100.00
Products				
H_2		497		0.07
C_1		7,585		1.15
C_2		7,549		1.15
$C_2^=$		5,187		0.79
Total C_2 and lighter		20,818		3.16
H_2S		1,032		0.16
C_3	2,090	15,262	4.18	2.32
$C_3^=$	4,027	30,504	8.05	4.63
IC_4	2,064	16,958	4.13	2.57
NC_4	1,827	15,579	3.65	2.36
$C_4^=$	4,178	37,150	8.36	5.64
Gasoline (C_5+ to 430°F TBP)	28,597	312,073	57.19	47.37
LCO (430°F TBP to 650°F TBP)	9,435	126,004	18.87	19.13
DO (650°F+ TBP)	3,718	55,994	7.44	8.50
Coke		27,440		4.17
Total	55,936	658,814	111.87	100.00
True Conversion			73.7	72.3
Inerts		5,143		

5.2 HEAT BALANCE

A cat cracker continually adjusts itself to stay in heat balance. This means that the reactor and regenerator heat flows must be equal (Figure 5-4). Simply stated, the unit produces and burns enough coke to provide energy to:

- increase the temperature of the fresh feed, recycle, and atomizing steam from their preheated states to the reactor temperature.
- provide the endothermic heat of cracking.
- increase the temperature of the combustion air from the blower discharge temperature to the regenerator flue gas temperature.
- make up for heat losses from the reactor and regenerator to surroundings.
- provide for miscellaneous heat sinks, such as stripping steam and catalyst cooling.

Heat balance can be performed around the reactor, around the stripper-regenerator, and as an overall heat balance around the reactor-regenerator. The stripper-regenerator heat balance can be used to calculate catalyst circulation rate and thus catalyst-to-oil ratio.

5.2.1 Heat Balance Around Stripper-Regenerator

If a reliable spent catalyst temperature is not available, the stripper is included in the heat balance envelope (II) as shown in Figure 5-4. The combustion of coke in the regenerator accounts for the following heat requirements:

- Heat to raise air from the blower discharge temperature to the regenerator dense phase temperature.
- Heat to desorb the coke from the spent catalyst.
- Heat to raise the temperature of the stripping steam to the reactor temperature.
- Heat to raise the coke on the catalyst from the reactor temperature to the regenerator dense phase temperature.
- Heat to raise the coke products from the regenerator dense temperature to flue gas temperature.
- Heat to compensate for regenerator heat losses.
- Heat to raise the spent catalyst from the reactor temperature to the regenerator dense phase.

Using the operating data provided in the case study, Example 5-5 shows heat balance calculations around the stripper-regenerator. As shown, the heat balance results are used to determine the catalyst circulation rate and the delta coke. Delta coke is the difference between coke on the spent catalyst and coke on the regenerated catalyst.

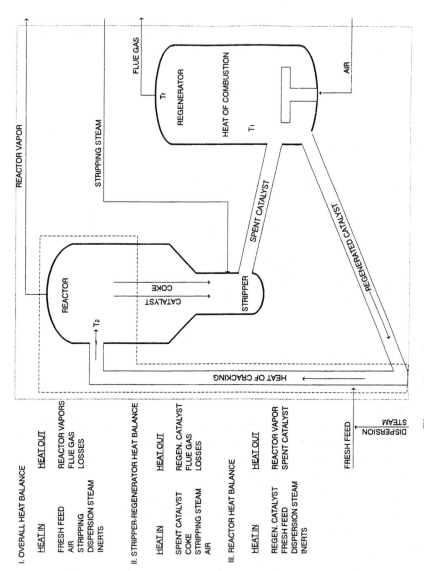

Figure 5-4. Reactor-regenerator heat balance.

Example 5-5
Stripper-Regenerator Heat Balance Calculations

I. Heat generated in the regenerator:

C to CO_2 = 24,323 lbs/hr × 14,087 Btu/lb = 342.6 × 10^6 Btu/hr

H_2 to H_2O = 2,696 lbs/hr × 51,571 Btu/lb = 139.0 × 10^6 Btu/hr

S to SO_2 = 212 lbs/hr × 3,983 Btu/lb = 0.84 × 10^6 Btu/hr

Total heat released in the regenerator:

342.6 + 139 + 0.84 = 482.4 × 10^6 Btu/hr

II. Required heat to increase air temperature from blower discharge to the regenerator dense phase temperature:
From Figure 5-5, enthalpies of air at 374°F and at 1309°F are 90 Btu/lb and 355 Btu/lb. Therefore, the required heat is = 407,493 lbs/hr × (355 − 90) Btu/lb = 108.0 × 10^6 Btu/hr

III. Energy to desorb coke from the spent catalyst:
Desorption of coke = 27,231 lbs/hr × 1,450 Btu/lb = 39.5 × 10^6 Btu/hr

IV. Energy to heat the stripping steam:
Enthalpy of 50 psig-saturated steam = 1,179 Btu/lb
Enthalpy of 50 psig at 972°F = 1,519 Btu/lb
Change of enthalpy = 13,000 lbs/hr × (1,519 − 1,179) Btu/lb = 4.4 × 10^6 Btu/hr

V. Energy to heat the coke on the spent catalyst:
27,231 lbs/hr × 0.4 Btu/lb-°F × (1309 − 972)°F = 3.7 × 10^6 Btu/hr

VI. Energy to heat the flue gas from regenerator dense phase to regenerator flue gas temperature:
From Figure 5-5, enthalpy of flue gas at 1309°F = 365 Btu/lb and at 1330°F = 370 Btu/lb. The required heat is therefore = 433,445 lbs/hr × (370 − 355)°F = 2.6 × 10^6 Btu/hr

VII. Heat loss to surroundings:
assume heat loss from the stripper-regenerator (due to radiation and convection) is 4% of total heat of combustion, i.e., 0.04 × 482.4 MM Btu/hr = 19.3 × 10^6 Btu/hr

VIII. Energy required to heat the spent catalyst from its reactor to the regenerator temperature =

$$482.4 - 108.0 - 39.5 - 4.4 - 3.7 - 2.6 - 19.3 = 304.9 \times 10^6 \text{ Btu/hr}$$

IX. Calculation of catalyst circulation

$$\text{Catalyst Circulation} = \frac{304.9 \times 10^6 \text{ Btu/hr}}{(0.285 \text{ Btu/}°F\text{-lb}) \times (1309 - 972)°F}$$

$$= 3.178 \times 10^6 \text{ lbs/hr} = 26.4 \text{ short tons/min.}$$

Where: 0.285 is the catalyst heat capacity (see Figure 5-6)

cat/oil ratio = $3.178 \times 10^6 / 658,914 = 4.8$

$$\Delta \text{ Coke} = \frac{\text{Coke Yield, wt\%}}{\text{cat/oil ratio}} = \frac{4.2}{4.8} = 0.87 \text{ wt\%}$$

5.2.2 Reactor Heat Balance

The hot-regenerated catalyst supplies the bulk of the heat required to vaporize the liquid feed (and any recycle), to provide the overall endothermic heat of cracking, and to raise the temperature of dispersion steam and inert gases to the reactor temperature.

Heat In	Heat Out
Fresh Feed	Reactor Vapors
Recycle	Flue Gas
Air	Losses
Steam	

The calculation of heat balance around the reactor is illustrated in Example 5-6. As shown, the reactor heat balance is converged to determine the overall endothermic heat of reaction. This approach to determining the heat of reaction is quite acceptable for daily unit monitoring. However, in designing a new cat cracker, it is often

(text continued on page 161)

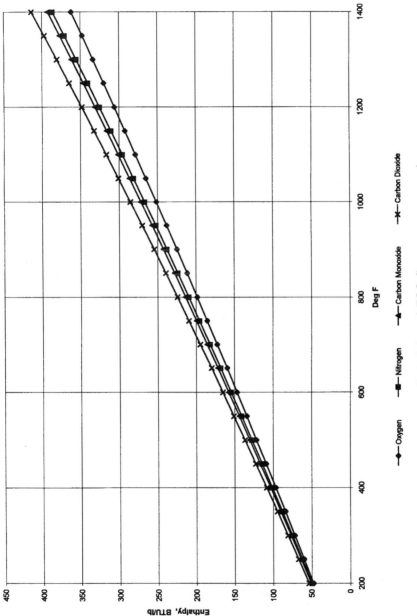

Figure 5-5. Enthalpies of FCC flue gas components.

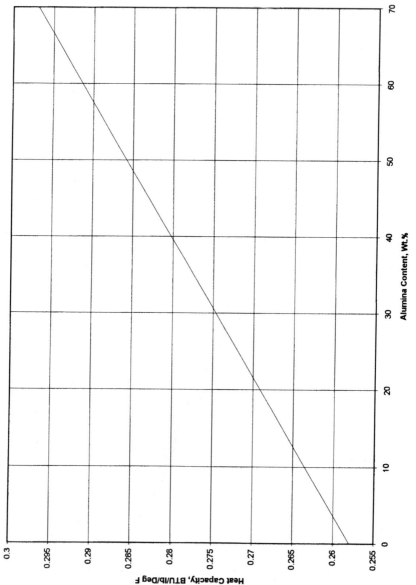

Figure 5-6.   Heat capacity of the FCC catalyst as a function of the catalyst's alumina content.

(text continued from page 158)

necessary to have a correlation that calculates the heat of reaction to determine other operating parameters, such as preheat temperature. Depending on conversion level, catalyst type, and feed quality, the heat of reaction can vary from 120 Btu/lb to 220 Btu/lb. Heat of reaction is a useful tool and indirect indication of heat balance accuracy. Monitoring the heat of reaction on a regular basis provides insight into various types of reactions occurring in the riser.

Example 5-6
Reactor Heat Balance

I. Heat into the Reactor
   1. Heat with Regenerator Catalyst = 3.178×10^6 lbs/hr $\times$ 0.285 Btu/lb-°F $\times$ 1309°F = $1,185.6 \times 10^6$ Btu/hr.
   2. Heat with the fresh feed:
      At feed temperature of 594°F, °API gravity = 25.2 and K factor = 12.08, the feed liquid enthalpy is 405 Btu/lb (see Figure 5-7), therefore, heat content of the feed is = 658,914 lbs/hr $\times$ 405 Btu/lb = 266.9×10^6 Btu/hr.
   3. Heat with atomizing steam:
      From steam tables, enthalpy of 150# saturated steam = 1176 Btu/lb, therefore, heat with steam = 10,000 lbs/hr $\times$ 1176 Btu/lb = 11.8×10^6 Btu/hr.
   4. Heat of adsorption:
      The adsorption of coke on the catalyst is an exothermic process; the heat associated with this adsorption is assumed to be the same as desorption of coke in the regenerator, i.e., 35.3×10^6 Btu/hr.
   Total heat in = 1,185.6 + 266.9 + 11.8 + 35.3 = $1,499.6 \times 10^6$ Btu/hr.
II. Heat Out of the Reactor
   1. Heat with spent catalyst = 3178×10^6 lbs/hr $\times$ 0.285 Btu/lb-°F $\times$ 972°F = 880.4×10^6 Btu/hr.
   2. Heat required to vaporize feed:
      From Figure 5-8, enthalpy of reactor vapors = 778 Btu/lb, therefore, heat content of the vaporized products = 658,814 lbs/hr $\times$ 778 Btu/lb = 512.6×10^6 Btu/hr.
   3. Heat content of steam:
      Enthalpy of steam @ 972°F = 1519 Btu/lb, therefore, heat content of steam = 10,000 lbs/hr $\times$ 1,519 Btu/lb = 15.2 M Btu/hr.
   4. Heat loss to surroundings:
      Assume heat loss due to radiant and convection to be 2% of heat with the regenerated catalyst, i.e., $0.02 \times 304.9 = 6.1 \times 10^6$ Btu/hr.

III. Calculation of Heat of Reaction
    Total heat out = Total heat in
    Total heat out = 880.4 + 512.6 + 15.2 + 6.1 + overall heat of reaction
    Total heat in = $1,499.6 \times 10^6$ Btu/hr
    Overall Endothermic Heat of Reaction = 85.3×10^6 Btu/hr or →
    129 Btu/lb of feed

5.2.3 Analysis of Results

Having performed the material and heat balances, the final step in a test run is discussion of the key findings and any recommendations to improve unit operation. The discussion should focus on factors affecting product quality and any abnormal results.

In the previous examples, the feed characterization correlations in Chapter 2 are used to determine composition of the feedstock. The results show that the feedstock is predominantly paraffinic, i.e., 61.6% paraffins, 19.9% naphthenes, and 18.5% aromatics. Paraffinic feedstocks normally yield the most gasoline with the least octane. This confirms the relatively high FCC gasoline yield and low octane observed in the test run. Of course, other factors such as the type of catalyst and operating parameters will also greatly affect the yield structure.

The coke calculation showed the hydrogen content to be 9.9 wt%. As discussed in Chapter 1, every effort should be made to minimize the hydrogen content of the coke entering the regenerator. The hydrogen content of a well-stripped catalyst is in the range of 5 to 6 wt%. A 9.9 wt% hydrogen in coke indicates either poor stripper operation and/or erroneous flue gas analysis.

5.3 PRESSURE BALANCE

Pressure balance deals with the hydraulics of catalyst circulation in the reactor/regenerator circuit. Pressure balance is about the various pressure increases and decreases in the circuit that affect the available differential pressure across the slide/plug valves, air blower, and wet gas compressor loadings.

A clear understanding of the pressure balance is extremely important in squeezing the most out of a unit. The incremental capacity could come from increased catalyst circulation or from altering the differential pressure between reactor-regenerator to "free up" the wet

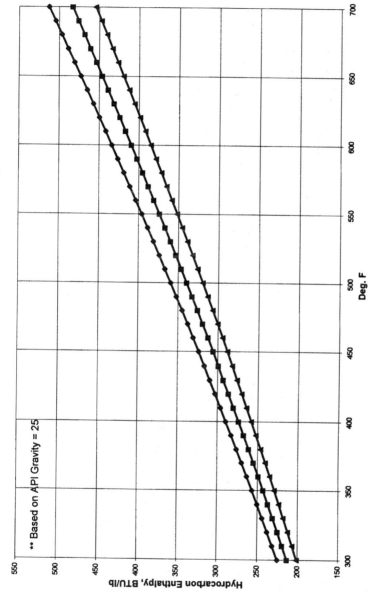

Figure 5-7. Hydrocarbon liquid enthalpies at various Watson K factors.

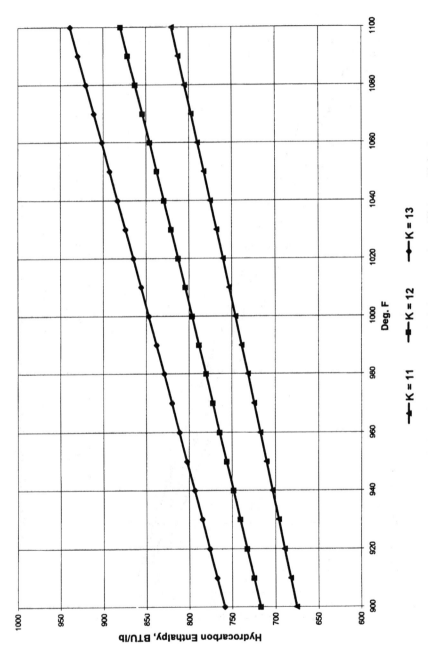

Figure 5-8. Hydrocarbon vapor enthalpies at various Watson K factors.

gas compressor or air blower loads. One must know how to manipulate pressure balance to identify the "true" constraints of the unit.

Using the drawing(s) of the reactor-regenerator, the unit engineer must be able to go through the pressure balance and determine whether it makes sense. He or she needs to calculate and estimate pressures, densities, pressure buildup in the standpipes, etc. The potential for improvements could be substantial.

5.3.1 Basic Fluidization Principals

A fluidized catalyst behaves like an aerated water and thus exhibits liquid-like characteristics. For example, flow occurs in the direction of lower density, or the difference in pressure between any two points in a bed is equal to the static head of the bed between these points.

FCC catalyst can only be made to flow like a liquid if the pressure force is transmitted through the catalyst particles and not the vessel wall. This means that the catalyst must always remain in a fluidized state as it makes a loop around the reactor/regenerator.

To illustrate the application of the above principals in FCC operations, the role of each major component of the circulation circuit is discussed in the following sections, followed by an actual case study. As a reference, Appendix 8 contains fluidization terms and definitions commonly used in the FCC.

5.3.2 Major Components of Reactor-Regenerator Circuit

The major components of the reactor-regenerator circulation circuit that either produce or consume pressure are as follows:

- Regenerator catalyst hopper
- Regenerated catalyst standpipe
- Regenerated catalyst slide or plug valve
- Riser
- Reactor-stripper
- Spent catalyst standpipe
- Spent catalyst slide, or plug, valve

Regenerator catalyst hopper

In some FCC units, the regenerated catalyst flows through a hopper prior to entering the standpipe. The hopper, being internal to the regenerator and often of an inverted cone design, provides sufficient time for the regenerated catalyst to be deaerated before entering the standpipe. This way, the catalyst entering the standpipe will have a maximum flowing density. The higher the density, the greater pressure buildup in the standpipe.

Regenerated catalyst standpipe

The standpipe's length and height provides the driving force for transferring the catalyst from the regenerator to the reactor. The elevation difference between the standpipe entrance and the slide (plug) valve is the source of this pressure buildup. For example, if the height difference is 30 feet and the catalyst density is 40 lbs/ft^3, the pressure buildup is:

$$\text{Pressure gain} = 30 \text{ ft} \times \frac{40 \text{ lbs}}{\text{ft}^3} \times \frac{1 \text{ ft}^2}{144 \text{ in}^2} = 8.3 \text{ psi}$$

The key to obtaining a maximum pressure gain is to avoid defluidization of the catalyst over the length of the standpipe. Longer standpipes may require external aeration to compensate for the compression of the entrained gas as it travels down the standpipe. Aeration should evenly be added along the length of the standpipe. However, in most cases, sufficient flue gas is carried down with the regenerated catalyst to keep it fluidized; therefore, a supplemental aeration medium is unnecessary, and often, over-aeration leads to unstable catalyst flow. Aside from proper aeration, the flowing catalyst must contain sufficient 0-40 micron fines to avoid defluidization.

Regenerated catalyst slide valve

The purpose of the regenerated catalyst slide (plug) valve is threefold: to regulate the flow of the regenerated catalyst to the riser, to maintain the pressure head in the standpipe, and to protect the regenerator from a flow reversal. Associated with this control and protection is usually a 2 to 8 psi pressure drop across the valve.

Riser

The hot-regenerated catalyst is transported up the vertical riser and into the reactor-stripper. The driving force to carry this mixture of catalyst and vapors comes from a higher pressure at the base of the riser and a large density difference between the fluid leaving the slide valve and the mixture of cracked hydrocarbon vapors and catalyst. As for the pressure balance, this transport of catalyst results in a pressure drop in a range of 5 psi to 9 psi. This drop is to a large extent due to the static head and to a lesser extent friction and acceleration. In an existing riser, any operating changes such as higher catalyst circulation or lower vapor velocity that can affect density of reaction mixture can increase the pressure drop and thus affect the slide valve differential and percent opening.

Reactor-Stripper

In the reactor-stripper, a catalyst bed is constantly maintained for three important reasons: to provide enough residence time for proper stripping of the entrained hydrocarbon vapors prior to entering the regenerator, to provide an adequate static head for a controlled outflow of the spent catalyst to the regenerator, and to provide a sufficient backpressure to prevent reversal of a hot flue gas into the reactor system. Assuming a stripper with a 20 ft bed level and a catalyst density of 40 lbs/ft^3, the static pressure is:

$$20 \text{ ft} \times \frac{40 \text{ lbs/ft}^3}{144 \text{ in}^2/\text{ft}^2} = 5.5 \text{ psi}$$

Spent catalyst standpipe

From the bottom of the stripper, the spent catalyst flows into the spent catalyst standpipe. Sometimes the catalyst is partially defluidized in the stripper cone. To counter this, "dry" steam is usually added (through a distributor) to fluidize the catalyst prior to its entering the standpipe. The loss of fluidization in the stripper cone can cause a buildup of dense phase catalyst along the cone walls. This buildup can restrict catalyst flow into the standpipe, causing erratic flow and reducing pressure buildup in the standpipe.

Like the regenerated catalyst standpipe, the spent catalyst standpipe may require supplemental aeration to obtain optimum flow characteristics. "Dry" steam is usually used as an aeration medium.

Spent catalyst slide or plug valve

The spent catalyst slide valve is located at the base of the standpipe. Its function is to control the stripper bed level and regulate the flow of the spent catalyst into the regenerator. As with the regenerated catalyst slide valve, maintaining a catalyst level in the stripper will be at the expense of consuming a pressure differential in the range of 3 to 6 psi.

5.3.3 Case Study

To provide a step-by-step method of calculating the pressure balance, assume a survey of the reactor-regenerator circuit of a 50,000 bpd cat cracker produced the results as shown below:

Reactor dilute phase (dome) pressure	= 19.0 psig
Reactor catalyst "dilute phase" bed level	= 25.0 ft
Reactor-Stripper catalyst bed level	= 18.0 ft
Reactor-Stripper catalyst density	= 40 lbs/ft³
Spent catalyst standpipe elevation	= 14.4 ft
Pressure above the spent catalyst slide valve	= 26.1 psig
Spent catalyst slide valve ΔP (@ 55 % opening)	= 4.0 psi
Regenerator "dilute phase" catalyst level	= 27.0 ft
Regenerator "dense phase" catalyst bed level	= 15.0 ft
Catalyst density in the regenerator "dense phase"	= 25 lbs/ft³
Regenerated catalyst standpipe elevation	= 30.0 ft
Pressure above the regenerated catalyst slide valve	= 30.5 psig
Regenerated catalyst slide valve ΔP (@ 30% opening)	= 5.5 psi
Reactor-regenerator pressure ΔP	= 3.0 psi

See also Figure 5-9 for a graphical representation of the preliminary results.

Starting with the reactor dilute pressure as the working point, the pressure head corresponding to 25 feet of dilute catalyst fines is:

$$(25 \text{ ft}) \times (0.6 \text{ lb/ft}^3) \times (1 \text{ ft}^2/144 \text{ in}^2) = 0.1 \text{ psig}$$

Therefore, the pressure at the top of the stripper bed is:

$$19.0 + 0.1 = 19.1 \text{ psig}$$

The static-pressure head in the stripper is:

$$(18 \text{ ft}) \times (40 \text{ lbs/ft}^3) \times (1 \text{ ft}^2/144 \text{ in}^2) = 5.0 \text{ psig}$$

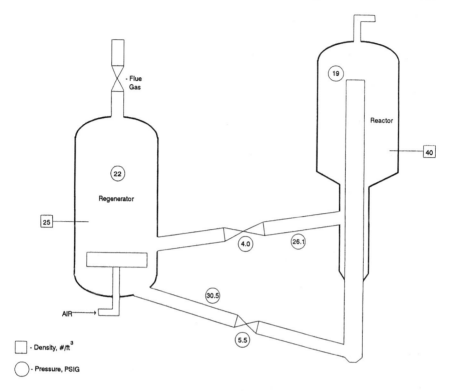

Figure 5-9. Preliminary findings of the pressure balance survey.

The pressure above the spent catalyst standpipe is:

19.1 + 5.0 = 24.1 psig

The pressure buildup in the spent catalyst standpipe is:

26.1 − 24.1 = 2 psi

The pressure below the spent catalyst slide valve is:

26.1 − 4.0 = 22.1 psig

The pressure head corresponding to 28 feet of dilute catalyst fines in the regenerator is:

(28 ft) × (1 lb/ft³) × (1 ft²/144 in²) = 0.2 psig

The pressure in the regenerator dome is:

22.1 − 0.2 = 21.9 psig

The static pressure head in the regenerator is:

(18 ft) × (25 lb/ft³) × (1 ft²/144 in²) = 3.1 psig

The pressure above the regenerated catalyst standpipe is:

22.1 + 3.1 = 25.2 psig

The pressure buildup in the regenerated catalyst standpipe is:

30.5 − 25.2 = 5.3 psi

The pressure below the regenerated catalyst slide valve is:

30.5 − 5.5 = 25 psig

The pressure drop in the vertical riser is:

25 − 19 = 6 psi

The catalyst density in the spent catalyst standpipe is:

(2.0 lbs/in²) × (144 in²/ft²)/(14.4 ft) = 20 lbs/ft³

The catalyst density in the regenerated catalyst standpipe is:

(5.3 lbs/in²) × (144 in²/ft²)/(30 ft) = 25.4 lbs/ft³

Figure 5-10 shows the results of the above pressure balance survey.

Analysis of the findings

The results of the pressure balance survey indicate that neither the spent nor the regenerated catalyst standpipe is generating "optimum" pressure head. This is evidenced by the low catalyst densities of 20 lbs/ft³

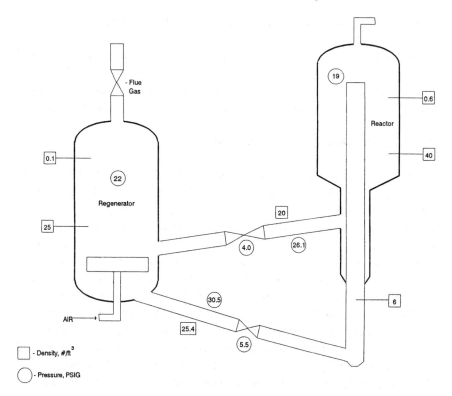

Figure 5-10. Results of the pressure balance survey showing standpipe calculated densities.

and 25.4 lbs/ft³, respectively. As indicated in Chapter 8, there are several factors causing the low pressure buildup, including "under" or "over" aeration of the standpipes. In a well-fluidized standpipe, the expected catalyst density is in the range of 35–45 lbs/ft³.

If the catalyst density in the spent catalyst standpipe were 40 lbs/ft³ instead of 20 lbs/ft³, the pressure buildup would have been 4.0 psi instead of 2.0 psi. The extra 2 psi can be used to circulate more catalyst and possibly to lower the reactor pressure (if a high rate of compressor "kickback" is used to maintain the accumulator pressure); its rate can be minimized.

In the regenerated catalyst standpipe, a 40 lbs/ft³ catalyst density vs. a 25.4 lbs/ft³ density produces 3 psi more pressure head, again allowing an increase in circulation or reducing the regenerator pressure (gaining more combustion air).

5.4 PROCESS CONTROL INSTRUMENTATION

The purpose of process control instrumentation is to allow the FCC unit to operate continuously in a safe, monitored mode with limited board operator intervention. There are two types of process control: "basic supervisory" control and "advanced" process control (APC).

5.4.1 Basic Supervisory Control

The primary controls in the reactor-regenerator section are flow, temperature, pressure, and catalyst level. The flow controllers are often used to set desired flows for the fresh feed, stripping steam, and dispersion steam. Each flow controller usually has three modes of control: *manual, auto,* and *cascade.* In manual mode, the operator manually opens or closes a valve to the desired percent opening. In auto mode, the operator enters the desired flow rate as a set-point. In cascade mode, the controller set-point is an input from another controller.

The reactor temperature is controlled by a temperature controller which regulates the regenerated catalyst slide valve. The control of regenerator temperature depends on its mode of operation. In partial combustion, the regenerator temperature is controlled by adjusting the flow of combustion air to the regenerator. In full burn, regenerator temperature is a function of operating conditions such as reactor temperature and slurry recycle.

The reactor pressure is not directly controlled; instead, it floats on the main column overhead receiver. A pressure controller on the overhead receiver indirectly controls the reactor pressure. The regenerator pressure is often controlled directly by regulating the flue gas slide or butterfly valve. In some cases, the flue gas slide or butterfly valve is used to control the differential pressure between regenerator and reactor.

The reactor or stripper catalyst level is controlled with a level controller which regulates the movement of the spent catalyst slide valve. The regenerator level is manually controlled to maintain catalyst inventory of the unit.

Regenerated and spent catalyst slide valve
low differential pressure override

Normally, the reactor temperature and the reactor bed level controllers regulate the movement of the regenerated and spent catalyst slide valves. The algorithm of these controllers can drive the valves either fully open or fully closed if the controller setpoint is unobtainable. It is extremely important that a positive and stable pressure differential be maintained across both the regenerated and spent catalyst slide valves. For these reasons, a "low differential pressure controller" is required to override the temperature /level controllers should these valves open too much.

The direction of catalyst flow around the reactor/regenerator must always be from the regenerator to the reactor and from the reactor back to the regenerator. A negative differential pressure across the regenerated catalyst slide valve would allow fresh feed and oil-soaked catalyst to back-flow from the riser into the regenerator. This "flow reversal" can result in an uncontrolled afterburn and possible equipment damage. A negative pressure differential across the spent catalyst slide valve would allow air to back-flow from the regenerator and into the reactor.

To protect the reactor-regenerator against the possibility of a flow reversal, pressure differential controllers are used to monitor and control the differential pressure across the regenerated and spent catalyst slide valves. If the differential pressure across the slide valve falls below a minimum set-point, the control of the valve is automatically transferred to the pressure differential controller (PDIC). Only after the PDIC set-point is satisfied will the control of the slide valve return to either reactor temperature or reactor level.

5.4.2 Advance Process Control

To maximize the FCC unit's profit, one must operate the unit simultaneously against as many constraints as possible. Examples of these constraints are limits in the air blower, wet gas compressor, reactor/regenerator temperatures, slide valve differential pressure, etc. The conventional regulatory control can only "push" the unit against one single constraint at a time. To overcome this limitation, a number of refiners have installed an advance process control (APC) package either within their DCS framework or as a host-based multivariable control.

The primary advantages of an APC are:

- It provides more precise control of the operating variables against the unit's constraints and therefore obtains incremental throughput or cracking severity.

- It is able to respond quickly to ambient disturbances such as cold fronts or rainstorms.
- It pushes against two or more constraints rather than one single constraint, e.g., maximum air blower and wet gas compressor capacities.

As mentioned above, there are two options for installing an APC in an existing cat cracker. One option is to install an APC within the DCS framework, and the other is to install a multivariable modeling/control package in a host-based computer. There are advantages and disadvantages of each package as indicated below.

Advantages of multivariable modeling and control

The multivariable modeling/control package is able to hold more tightly against constraints and recover more quickly from disturbances. This results in an incremental amount of capacity over a DCS-based APC. In APC on DCS framework, the control structure has to be designed, configured, and programmed for each specific unit. With a multivariable modeling/control package, the controlled and manipulated variables must be chosen, and their particular unit responses must be modeled. Programming and/or software configuring is typically not required here.

Disadvantages of multivariable modeling and control

A multivariable model is like a "black box"; the constraint measurements go in and the manipulated variables come out. Operators do not have a secure knowledge of what is tied to what. The DCS-based APC is installed in a modular form, meaning operators can understand what the controlled variable is tied to a little more easily. In addition, a multivariable model relies on a host computer; the host computer may have its own problems, such as may occur with computer-to-computer data links.

In any APC, the operator has to be educated and bought into it before he or she can use it. Of course, the control has to be properly designed, meaning that in the case of DCS-based control, the module has to be configured and "tuned" properly. With multivariable model APC, the model has to be built properly, with correct variables chosen for the inputs.

SUMMARY

In conclusion, the only proper way to evaluate the performance of a cat cracker is by conducting material and heat balance tests. Predictions and evaluation of changes in feedstocks, catalyst, operating parameters, and any mechanical hardware can be made by carrying out these tests. As discussed in the next chapter, material and heat balances are the foundations for determining the effects of operating variables on unit operation.

The material balance test run provides a standard and consistent approach for daily monitoring of a cat cracker. It also allows for accurate analysis of yields and proper comparison of unit performance. The composition of the products leaving the reaction section can be determined either through direct sampling of the reactor overhead line or conducting a unit test run. Each sampling technique has advantages and disadvantages.

The heat balance exercise provides a tool for in-depth insight into the operation of a cat cracker. Heat balance surveys are used to determine catalyst circulation rate, delta coke, and heat of reaction. The procedures described in this chapter can be easily programmed into a spreadsheet computer program to calculate material and heat balances on a routine basis.

The pressure balance provides an insight into the hydraulics of the catalyst circulation. Performing pressure balance surveys will help the unit engineer to recognize circulation "pinch points" and also to get the most out of the common constraints: the air blower and the wet gas compressor.

Finally, process control features control schemes to operate the unit smoothly and safely. In conjunction with that, implementation of an APC package, whether within the DCS framework or as host-based multivariable control, provides more precise control of operating variables against the unit's constraints and, therefore, obtains incremental throughput or cracking severity.

REFERENCES

1. Davison Div., W. R. Grace & Co., "Cat Cracker Heat and Material Balance Calculations," Grace Davison *Catalagram,* No. 59, 1980.
2. Hsieh, C. R. and English, A. R., "Two Sampling Techniques Accurately Evaluate Fluid-Cat-Cracking Products," *Oil & Gas Journal,* June 23, 1986, pp. 38–43.

CHAPTER 6

Products
and Economics

The topics and discussions in the previous chapters are important in understanding the operation of a cat cracker. However, the purpose of the FCC unit is to maximize profitability for the refinery. The cat cracker provides the basic economics to make the refinery a viable entity. Over the years, refineries without cat crackers have been shut down because they were not profitable.

It is as important to understand FCC economics as it is to understand the heat or pressure balance in the FCC unit. The dynamics of FCC economics changes rapidly and is dependant on market conditions and the availability of "quality" feedstocks.

The objective of this chapter is to discuss the factors affecting yields and qualities of FCC product streams. The 1990 Clean Air Act Amendment (CAAA) has imposed greater restrictions on quality standards for gasoline and diesel. FCC, being the major contributor to the gasoline and diesel pool, is significantly affected by these new regulations. The section on FCC economics describes several operating options that can be used to maximize FCC unit performance and, consequently, the refinery's profit margin.

6.1 FCC PRODUCTS

The cat cracker's function is to convert the less valuable gas oils to more valuable products. A major objective of most FCC units is to maximize the conversion of gas oil to gasoline and LPG. The products from the cat cracker are:

- Dry Gas
- LPG

- Gasoline
- LCO
- HCO
- Decanted Oil
- Coke

6.1.1 Dry Gas

The gas (C_2 and lighter) leaving the "sponge oil absorber" tower is commonly referred to as "dry gas"; it mainly contains hydrogen, methane, ethane, ethylene, and traces of hydrogen sulfide. Once the gas is amine-treated for removal of H_2S and other traces of acid gases, it is blended into the refinery fuel gas system. Depending on the volume percent of hydrogen in the dry gas, some refiners recover the hydrogen using processes such as cryogenics, pressure-swing absorption, or membrane separation. The recovered hydrogen is often used in hydrotreating.

Dry gas is an undesirable by-product of the FCC unit; its excessive yield loads up the wet gas compressor and is often a constraint in the cat cracker. The dry gas yield is primarily due to thermal cracking, metals in the feed, and nonselective catalytic cracking. In the FCCU, the main factors which contribute to the increase in production of dry gas are:

- Increase in the concentration of metals (nickel, vanadium, etc.) in the feed and on the catalyst.
- Increase in reactor or regenerator temperatures.
- Increase in the residence time of hydrocarbon vapors in the reactor.
- Decrease in the performance of the feed nozzles.
- Increase in the aromaticity of the feed.

6.1.2 LPG

The overhead stream from the debutanizer or stabilizer is usually referred to as LPG. The cat cracker's LPG is rich in the olefins propylene and butylene. These light olefins play an important role in the manufacture of reformulated gasoline. Depending on the refinery's configuration, the cat cracker's LPG is used in the following areas:

- *Chemical sale*, where the LPG is separated into C_3's and C_4's. The C_3's are sold as refinery or chemical grade propylene. The C_4 olefins are polymerized.
- *Direct blending*, where the C_4's are blended into the refinery's gasoline pool to regulate vapor pressure and to enhance the octane number. However, new gasoline regulations require reduction of the vapor pressure, thus displacing a large volume of C_4's for alternative uses.
- *Alkylation*, where the olefins are reacted with isobutane to make a very desirable gasoline blending stock. Alkylate is an attractive blending component because it has no aromatics or sulfur, low vapor pressure, low end point, and high research and motor octane ratings.
- *MTBE*, where isobutylene is reacted with methanol to produce an oxygenate gasoline additive called methyl tertiary butyl ether (MTBE). MTBE is added to gasoline to meet the minimum oxygen requirement for "reformulated" gasoline.

The LPG yield and its olefinicity can be increased by:

- Changing to a catalyst which minimizes "hydrogen transfer" reactions.
- Increasing the conversion.
- Decreasing cracking residence time, particularly the amount of time product vapors spend in the reactor housing before entering the main column.
- Adding ZSM-5.

An FCC catalyst containing zeolite with a low hydrogen transfer rate reduces the resaturation of the produced olefins in the riser. As stated in Chapter 4, primary cracking products in the riser are highly olefinic. Most of these olefins are in the gasoline boiling range; the rest appear in the LPG and LCO boiling range. The LPG olefins do not crack further, however, they can become saturated by hydrogen transfer. The gasoline and LCO range olefins can be cracked again to form gasoline range olefins and LPG olefins. The olefins in the gasoline and LCO range can also cyclize to form cycloparaffins. The cycloparaffins can react through H_2 transfer with olefins in the LPG and gasoline to produce aromatics and paraffins. Therefore, a catalyst which inhibits hydrogen transfer reactions will increase olefinicity of the LPG.

The conversion increase is accomplished by manipulating the following operating conditions:

* *Increasing the reactor temperature.* Increasing the reactor temperature beyond the peak gasoline yield results in overcracking of the gasoline and LCO fractions. The rate of production and olefinicity of the LPG will increase.
* *Increasing feed/catalyst mix zone temperature.* Conversion and LPG yield can be increased by injecting a portion of the feed, or naphtha, at an intermediate point in the riser (see Figure 6-1). The splitting or segregation of the feed results in a high-mix zone temperature producing more LPG and more olefins. This practice is particularly useful in units where the reactor temperature is already maximized due to metallurgy constraint.

Reduction of the catalyst/hydrocarbon time in the riser, coupled with the elimination of post-riser cracking, reduces the saturation of the "already produced olefins" and allows the refiner to increase the reaction severity. These actions enhance the olefin yields and still

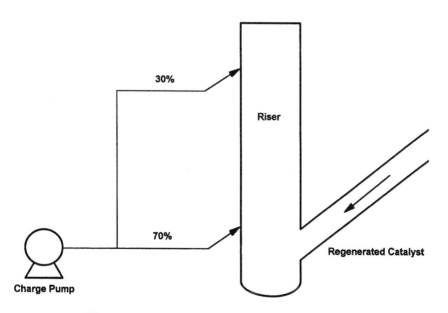

Figure 6-1. A typical feed segregation scheme.

operate within the constraint of the wet gas compressor. Elimination of post-riser residence time (direct connection of the reactor cyclones to the riser) virtually eliminates the undesired thermal and nonselective catalytic cracking and thus reduces dry gas and diolefin yields.

Adding ZSM-5 catalyst additive is another process available to the refiner to boost production of light olefins in the FCC unit. ZSM-5, at a typical concentration of 0.5 to 3.0 wt%, is usually used in a number of FCC units to increase the gasoline octane and light olefins. As part of the cracking of low octane components in the gasoline, ZSM-5 also makes C_3, C_4, and C_5 olefins (see Figure 6-2). Paraffinic feedstocks respond the most to ZSM-5.

6.1.3 Gasoline

The FCC gasoline has always been the most valuable product of a cat cracker unit. This gasoline accounts for about 35 vol% of the total U.S.-produced gasoline. Historically, the objective has been to achieve maximum gasoline yield with highest octane.

Gasoline Yield

For a given feedstock, the gasoline yield can be increased by the following actions:

• Increasing catalyst-to-oil ratio by decreasing the feed preheat temperature.
• Increasing catalyst activity by increasing fresh catalyst addition rate or fresh catalyst activity.
• Increasing gasoline end point by reducing the main column top pumparound rate.
• Increasing reactor temperature (if the increase does not overcrack the already produced gasoline).

Gasoline Quality

The Clean Air Act Amendment (CAAA) passed in November 1990 has set new quality standards for U.S. gasoline. A complete discussion of the new gasoline formulation requirements can be found in Chapter 9.

The key components affecting FCC gasoline quality are octane, benzene, and sulfur, and are discussed in the following sections.

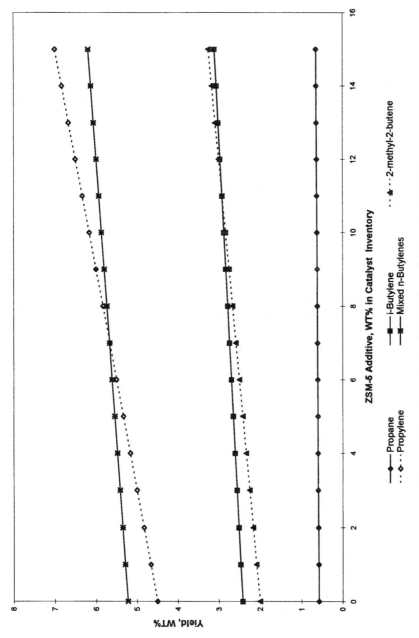

Figure 6-2.   The effect of ZSM-5 on light-ends yield [5].

Octane. An octane number is a quantative measure of a fuel mixture's resistance to "knocking." The octane number of a particular sample is measured against a standard blend of n-heptane, which has 0 octane, and iso-octane, which has 100 octane. The percent of iso-octane that produces the same "knock" intensity as the sample is reported as the octane number.

Two octane numbers are routinely used to simulate the engine performance: The *research octane number* (RON) simulates gasoline performance under low severity (@ 600 rpm and 120°F air temperature), whereas the *motor octane number* (MON) reflects more severe conditions (@ 900 rpm and 300°F air temperature). At the pump, *road octane*, which is the average of RON and MON, is reported.

The factors affecting gasoline octane are:

A. Operating Conditions
    1. *Reactor Temperature.* As a rule, an increase of 18°F in reactor temperature increases the RON by 1.0 number and MON by 0.4 number. However, the MON contribution comes from the aromatic content of the heavy end. Therefore, at high severity, the MON response to the reactor temperature can be greater than 0.4 number per 18°F.
    2. *Gasoline End Point.* The effect of gasoline end point on its octane number depends upon the feedstock quality and the severity of the operation. At low severity, lowering the end point will boost the MON. At high severity, lowering the end point will also lower the MON. Reducing the gasoline end point of a paraffinic feedstock may not impact the octane number; however, reducing gasoline end point produced from a naphthenic or an aromatic feedstock will lower the octane.
    3. *Gasoline Reid Vapor Pressure (RVP).* The RVP of the gasoline is regulated by varying the amount of C_4's, which increase gasoline octane. The octane response depends on whether the RVP increase is due to C_4 olefins or from butane. As a rule, the RON and MON gain 0.3 and 0.2 numbers for a 1.5 psi increase in RVP.
B. Feed Quality
    1. *°API Gravity.* The higher the °API gravity, the more paraffins in the feed, the lower the octane (Figure 6-3).
    2. *K Factor.* K factor is an indicator of feed paraffinicity, therefore, the higher the K factor, the lower the octane.

3. *Aniline Point.* Feeds with higher aniline point temperatures are less aromatic and more paraffinic, therefore, the higher the aniline point, the lower the octane.

4. *Sodium.* Additive sodium reduces unit conversion and loss of octane (Figure 6-4).

C. Catalyst

1. *Rare Earth.* Increasing the amount of rare earth on the zeolite decreases the octane (Figure 6-5).

2. *Unit Cell Size.* Decreasing the unit cell size increases octane (Figure 6-6).

3. *Matrix Activity.* Increasing the catalyst matrix activity increases the octane.

4. *Coke on the Catalyst.* Increasing the amount of coke on the regenerated catalyst lowers its activity and increase the octane.

Benzene. Most of the benzene in the gasoline pool comes from *reformate.* Reformate, the high-octane gasoline from a reformer unit, comprises about 30 vol% of the gasoline pool. Depending on the reformer feedstock and severity, reformate contains 3–5 vol% benzene.

The benzene content of the cat cracker gasoline is in a range of 0.5 to 1.3 volume percent. Since the FCC gasoline accounts for about 35 vol% of the gasoline pool, it is important to know which principal factors affect the cat cracker gasoline benzene levels. The benzene content in the FCC gasoline can be reduced by:

- Short contact time in the riser and in the reactor dilute phase.
- Lower cat-to-oil ratio and lower reactor temperature.
- A catalyst with a lower hydrogen transfer properties.

Sulfur. The major source of sulfur in the gasoline pool comes from the FCC gasoline. The sulfur level in gasoline is a strong function of the feed sulfur content (Figure 6-7). Hydrotreating the FCC feedstock reduces sulfur in the feedstock and consequently in the gasoline (Figure 6-8). Other factors which can lower sulfur content in gasoline are:

- Lower gasoline end point (see Figure 6-9).
- Lower reactor temperature (see Figure 6-10).
- Increased matrix activity of the catalyst.

(text continued on page 189)

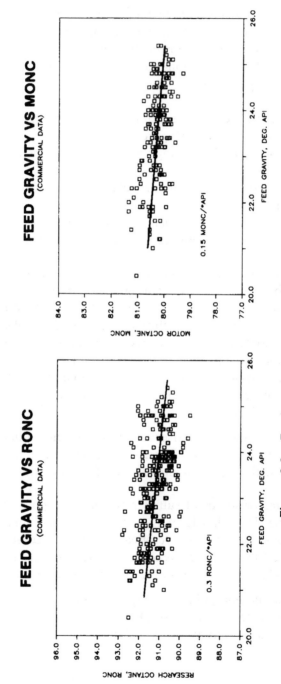

Figure 6-3. Feed gravity comparisons (MON and RON) [7].

RONC vs. SODIUM
COMMERICAL DATA

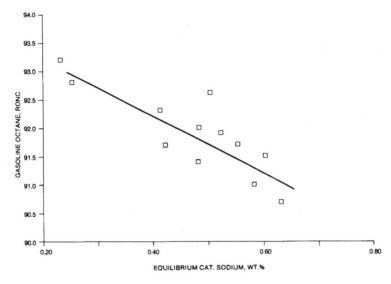

MONC vs. SODIUM
COMMERCIAL DATA

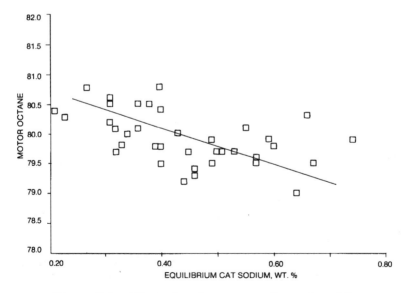

Figure 6-4. Effect of sodium on gasoline octane [8].

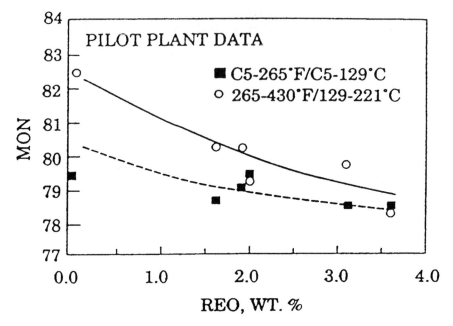

Figure 6-5. Effect of fresh REO on MON [9].

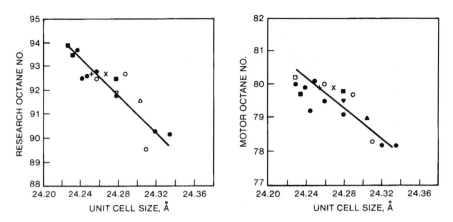

Figure 6-6. Effects of unit cell size on research and motor octane [10].

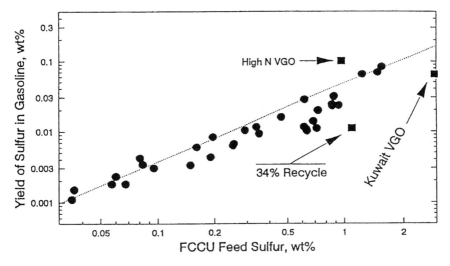

Figure 6-7. FCC gasoline sulfur yield [4].

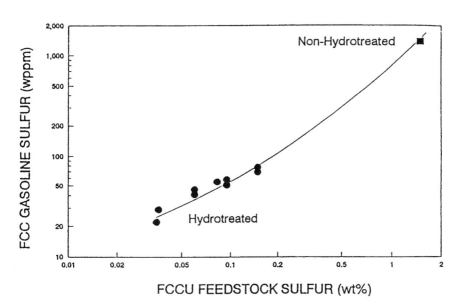

Figure 6-8. Hydrotreating reduces FCC gasoline sulfur [4].

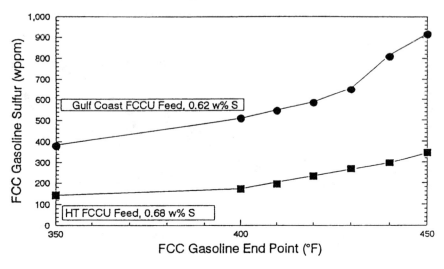

Figure 6-9.   FCC gasoline sulfur increases with end point [4].

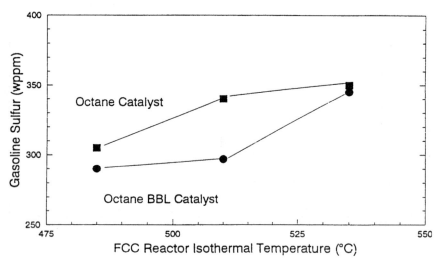

Figure 6-10.   FCC gasoline sulfur increases with temperature [4].

(text continued from page 183)

- Increase in the catalyst activity and hydrogen transfer properties.
- Increase in catalyst-to-oil ratio (Figure 6-11).

6.1.4 LCO

The emphasis to maximize gasoline yield has sometimes overshadowed the importance of other FCC products, particularly LCO. LCO is widely used as a blending stock in heating oil and diesel fuel. This is particularly important in winter when the value of light cycle oil can be greater than gasoline. Under these circumstances, many refiners adjust the FCC operation to increase LCO yield at the expense of gasoline.

LCO Yield

A refiner has several options to increase LCO yield. Since it is often desirable to maintain a maximum cracking severity while maximizing light cycle oil yield, the simplest way to increase LCO yield is to reduce the gasoline end point. Gasoline end point reduction is usually carried out by either increasing the main column top pumparound rate or increasing the top reflux rate.

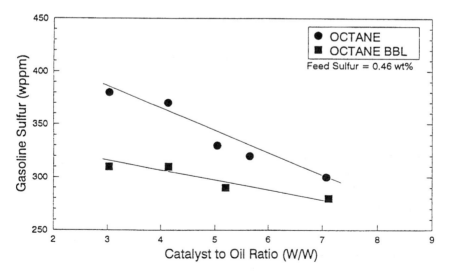

Figure 6-11. Increased catalyst-to-oil ratio decreases gasoline sulfur [4].

The LCO cut point is typically between 430°F and 650°F ASTM. Undercutting gasoline end point allows the heavy end of the gasoline fraction to be withdrawn with LCO, affecting only the apparent conversion and not causing changes in the flow rate of other products. Additionally, reducing the gasoline end point usually increases the octane because of the lower-octane-rated components which are present in the heavy end of gasoline.

Another method of increasing LCO yield is through the removal of the LCO boiling-range materials from the FCC feed. As shown in Table 6-1, the total refinery yield of products in the LCO boiling range will increase when the light ends are fractionated from the feed.

Some of the catalytic routes to maximize LCO yield are:

- Decrease in the reactor temperature.
- Decrease in the catalyst-to-oil ratio.
- Decrease in catalyst activity.
- Increase in HCO recycle.

LCO Quality

The CAAA of 1990 fuel standards for the new "over-the-road diesel fuel" require a maximum sulfur and aromatic concentration of 0.05 wt% (500 ppm) and 35 vol%, respectively. In addition, CARB has imposed

Table 6-1
Effects of Feed Fractionation on Total Distillate Yield

	Feedstock	
	"Raw" Gas Oil	"Fractionated" Gas Oil
Initial Boiling Point, °F	435	660
Final Boiling Point, °F	1,080	1,080
435°F to 660°F Content, wt%	8	0
Conversion, wt%	75.9	75.9
LCO, wt%	15.4	14.0
Potential FCC LCO, wt%	15.4	(0.92 × 14.0) = 12.9
Total Potential Refinery Distillate, wt%	15.4	(12.9 + 8.0) = 20.9

Source: Engelhard [6]

an additional restriction on diesel aromatics of 10 vol% for larger refineries (over 50,000 bpd capacity) and 20 vol% for smaller refineries. A minimum cetane number of 40 is also required. Since the aromatics content of a diesel fuel is a function of its cetane quality, the following section discusses cetane number and the principal factors affecting it.

Cetane. Like octane number, cetane number is a numerical indication of the ignition quality of a fuel such as kerosene, diesel, or heating oil. Unfortunately, any actions to increase gasoline octane will adversely affect LCO cetane quality. This is because the hydrocarbon types that produce a high-octane gasoline generally produce a low cetane number. For example, normal paraffinic hydrocarbons have a low octane number but a very high cetane number. Aromatics have a very low cetane number.

Cetane number is measured in a single-cylinder laboratory engine, whereas a more common indicator, cetane index, is a calculated value. Cetane index correlates adequately with the cetane number. Most refiners use the ASTM equation (Method D-976-80) to calculate the cetane index. The equation in Example 6-1 uses the 50% boiling point and °API gravity to calculate cetane index.

Example 6-1

Cetane Index Equation

$CI = 65.01 \ (\log T)^2 + 0.192 \ (°API) \times (\log T)$

$\qquad + .016 \ (°API)^2 - 0.0001809 \ (T)^2 - 420.34$

Where:

$\qquad CI = $ Cetane Index
$\qquad T = $ Mid boiling temperature, °F
$\qquad °API = $ °API Gravity at 60°F

For Example:

$\qquad T = 550°F$
$\qquad °API = 19$
$\qquad CI = 28.9$

Source: ASTM Standards, Method D-976-80

LCO is highly aromatic (50–75 wt%), and therefore it has a low cetane index (20–30). The cetane number and sulfur content of the LCO largely determine the amount of straight-run distillate which can be blended into the diesel or heating oil pool. Most of the aromatics in the LCO are di- and tri-aromatic (30–50 wt%) molecules. Hydrotreating the LCO can increase its cetane number. The degree of improvement depends on the severity of the hydrotreating. Mild hydrotreating (500–800 psig) can partially hydrogenate some of the di- and tri-aromatics and increase cetane by 1 to 5 number. Severe hydrotreating conditions (>1500 psig) can increase the cetane number above 40.

Other conditions which improve cetane are:

• Undercutting the FCC gasoline flow.
• Reducing the unit conversion.
• Using an "octane" catalyst.
• Processing paraffinic feedstock in the FCC unit.

6.1.5 HCO and Decant Oil

HCO is one of the sidecut streams from the main column that boils between LCO and the decanted oil. HCO is often used as a pump-around stream to transfer heat to the fresh feed and/or to the debutanizer reboiler. HCO is recycled to extinction, withdrawn as a product and processed into a hydrocraker, or blended with the decant oil.

Decant oil (DO) is the heaviest and often the lowest priced product from a cat cracker. DO is also called slurry oil, clarified oil, bottoms, and FCC residue. Depending on the refinery location and market availability, DO is typically blended into #6 fuel, sold as a carbon black feedstock, or even recycled to extinction.

The DO's yield depends largely on the quality of the feedstock and the conversion level. Naphthenic and aromatic feedstocks tend to yield more bottoms than a highly paraffinic feedstock. If the conversion is in the low to mid 70's, there are steps, such as increasing catalyst-to-oil ratio or using a catalyst with an active matrix, that a refiner can take to reduce slurry yield. However, if the conversion is in the 80's, there is little that can be done to reduce bottoms yields.

Decant Oil Quality

DO properties vary greatly depending on the FCC feedstock's quality and operating conditions. A better price is often obtained by selling the DO as carbon black feedstock (CBFS) than would be obtained by using it as a cutter stock. Aromaticity and ash content are the two most important properties of CBFS (Table 6-2).

°API gravity is a rough indication of aromaticity and boiling range. Aromaticity is generally measured by the Bureau of Mines Correlation Index (BMCI). To meet the CBFS specification, the DO's °API gravity must not exceed 2.0, and it should have a minimum BMCI of 120.

Performance of the reactor cyclones and the catalyst physical properties greatly affect the ash content of the decant oil. To meet the CBFS's ash requirement (maximum of 0.05 wt%), DO product may need to be filtered for the removal of the catalyst fines.

6.1.6 Coke

In a cat cracker, a portion of the feed is deposited on the catalyst as coke. Coke formation is a necessary by-product of the FCC operation; the heat released from burning the coke in the regenerator compensates for the heat lost in the riser.

Table 6-2
Typical Carbon Black Feedstock Specifications

Property	Specification
Gravity, °API	3.0, maximum
Asphaltenes, wt%	5.0, maximum
Viscosity, SSU @ 210°F	80, maximum
Sulfur, wt%	4.0, maximum
Ash, wt%	0.05, maximum
Sodium, ppm	15, maximum
Potassium, ppm	2, maximum
Flash, °F	200, minimum
BMCI*	120, minimum

Bureau of Mines Correlation Index (BMCI):
  $BMCI = (87,552/T) + [473.7 \times (141.5/131.6 + °API\ Gravity)] - 456.8$
Where:
  $T = Mid\ boiling\ point,\ °R$
For Example:
      $T = 710°F$
      $°API = 1.0$
      $BMCI = 123.9$

The structure and chemistry of coke formation is difficult to define. However, it is generally agreed that the types of coke in FCC come from at least four sources:

- *Catalytic coke* is a by-product of the cracking of FCC feed to lighter products. Its yield is mainly a function of conversion, catalyst type, and hydrocarbon/catalyst residence time in the reactor.
- *Contaminant coke* (metals coke) is produced by catalytic activity of metals such as nickel and vanadium deposited on the catalyst.
- *Feed residue coke* is the small portion of the feed which is directly deposited on the catalyst. This coke comes from the very heavy fraction of the feed and its yield is predicted by the Conradson or Ramsbottom carbon tests.
- *Catalyst circulation coke* is a "hydrogen-rich" coke from the reactor-stripper. Efficiency of catalyst stripping and the catalyst's pore size distribution affect the amount of the hydrocarbons carried over into the regenerator.

A proposed equation [1] to express coke yield is:

$$\text{Coke Yield, wt\%} = g\ (z_1, \ldots z_n) \times (C/O)^n \times (WHSV)^{n-1} \times [e^{(\Delta Ec/RTrx)}]$$

Where:

g	= function of feed quality, hydrocarbon partial pressure, catalyst type, CRC, etc.
n	= 0.65
C/O	= cat-to-oil ratio
WHSV	= weight hourly space velocity, weight of total feed/hr divided by weight of catalyst inventory in reaction zone, hr^{-1}
Δ_{Ec}	= activation energy ~ 2500 BTU/lb-mole
R	= gas constant, 1.987 BTU/lb-mole-°R
T_{Rx}	= reactor temperature, °R

The *coke yield* of a given cat cracker is essentially constant. The FCC produces enough coke to satisfy the heat balance. However, a more important term is *delta coke*. Delta coke is defined as the difference between the coke on the spent catalyst and the coke on the regenerated catalyst. At a given reactor temperature and constant CO_2/CO ratio, delta coke controls the regenerator temperature.

Reducing delta coke will lower the regenerator temperature. There are many benefits associated with a lower regenerator temperature. The higher cat/oil ratio improves product selectivity and/or provides the flexibility to process heavier feeds.

Many factors influence delta coke, including the design of the feed/catalyst injection system, riser design, operating conditions, and catalyst type. Following is a brief discussion of these factors:

- *Feed/Catalyst Injection.* A well-designed injection system provides a rapid and uniform vaporization of the liquid feed. This will lower delta coke by minimizing noncatalytic coke deposition as well as reducing the deposits of heavy material on the catalyst.
- *Riser Design.* A properly designed riser will help to reduce delta coke by reducing the back-mixing of already "coked-up" catalyst with fresh feed. This back-mixing causes unwanted secondary reactions.
- *Cat/Oil Ratio.* An increase in the cat/oil ratio reduces delta coke by spreading out some coke-producing feed components over more catalyst particles and thus lowering the concentration of coke on each particle.
- *Reactor Temperature.* An increase in the reactor temperature will also reduce delta coke by favoring cracking reactions over hydrogen transfer reactions. Hydrogen transfer reactions produce more coke than the cracking reactions.
- *Catalyst Activity.* An increase in catalyst activity will increase delta coke. As catalyst activity increases so does the number of adjacent active sites, which increases the tendency for hydrogen transfer reactions to occur. Hydrogen transfer reactions are bimolecular and require adjacent active sites.

6.2 FCC ECONOMICS

In most refineries, the cat cracker's operational philosophy is dictated by refinery economics. Economics of a refinery is divided into *internal* and *external* economics.

Internal economics depends largely on the cost of raw crude and the FCC unit's yields. The cost/benefit of the crude purchased often outweighs the debit that it may have on the cat cracker yields. Some refiners operate their units not so much by the economics but by a kind of intuition. For example, they may drive for more throughput, but this may not be the most profitable approach.

External economics are factors that are generally forced upon the refineries. Refiners prefer not to have their operations dictated by external economics. However, they may have to meet particular requirements such as those for reformulated gasoline.

To maximize the FCCU's profit, the unit must be operated against all its mechanical and operating constraints. Generally speaking, the incremental profit of increasing feed to the FCCU is more than the incremental profits obtained by increasing conversion. The general driving force is usually to maximize gasoline yield while maintaining the minimum octane requirements that meet gasoline specifications.

Because of the high cost of new grass root cat crackers and the importance of the FCC units on overall refinery profitability, improvements should be made to the existing units to maximize their performance. These performance indices are:

- Improving product selectivities
- Enhancing operating flexibility
- Increasing unit capacity
- Improving unit reliability
- Reducing operating cost
- Meeting product specifications
- Reducing emissions

Product selectivity simply means producing more liquid products and less coke and gas. Depending on the unit's objectives and constraints, below are some of the steps that directionally improve product selectivity:

- *Feed Injection.* An improved feed injection system provides optimum atomization and distribution of the feed for rapid mixing and complete vaporization. The benefits of improved feed injection are reduced coke deposition, reduced gas yield, and improved gasoline yield.
- *Riser Termination.* Good riser termination devices, such as closed cyclones, minimize the vapor and catalyst holdup time in the reactor vessel. This reduces unnecessary thermal and nonselective catalytic recracking of the reactor products. The benefits are a reduction in dry gas and a subsequent improvement in conversion, gasoline octane, and flexibility for processing marginal feeds.

- *Reactor Stripper.* Operational and hardware changes to the stripper design improve its performance by not admitting the entrained and adsorbed hydrocarbon in the spent catalyst into the regenerator. The benefits are lower delta coke and more liquid products.
- *Air and Spent Catalyst Distribution.* Modifications to the air and spent catalyst distributors permit uniform dispersion of air and spent catalyst into the regenerator. Improvements are lower carbon on the catalyst and less catalyst sintering. The benefits are a cleaner and higher-activity catalyst, which results in more liquid products and less coke and gas.

Examples of increasing operating flexibility are:

- *Processing Residue or "Purchased" Feedstocks.* Sometimes, the option of processing supplemental feed or other components such as atmospheric residue, vacuum residue, and lube oil extract is a means of increasing the yields of higher-value products and reducing the costs of raw material by purchasing less expensive feedstocks.
- *ZSM-5 Additive.* Seasonal or regular use of ZSM-5 catalyst will center-crack the low-octane paraffin fraction of the FCC gasoline. The results are increases in propylene, butylene, and octane—all at the expense of the FCC gasoline yield.
- *Catalyst Cooler(s).* Installing a catalyst cooler(s) is a way to control and vary regenerator heat removal and thus allow processing of a poor quality feedstock to achieve increased product selectivity.
- *Feed Segregation.* Split feed injection involves charging a portion of the same feed to a different point in the riser. This is another tool for increasing light olefins and boosting gasoline octane.

An example of increasing FCCU capacity is oxygen enrichment:

- *Oxygen Enrichment.* In a cat cracker which is either air blower or regenerator velocity limited, enrichment of the regenerator air can increase capacity or conversion, provided there is good air/catalyst distribution and that the extra oxygen does not just burn CO to CO^2.

In recent years, numerous mechanical improvements have been implemented to the FCCU to increase the run length and to minimize maintenance work during turnarounds. Examples are:

- *Expansion Joints.* Improvement in bellows metallurgy to Alloy 800H or Alloy 625 has reduced the failures caused by stress corrosion cracking induced by polythionic acid. Additionally, placing fiber packing in the bellow-to-sleeve annulus, instead of purging with steam, has reduced bellows cracking.
- *Slide or Plug Valves.* Cast-vibrating of the refractory lining and stem/guide purge modifications have minimized stress cracking and erosion.
- *Air Distributors.* Improvements in the metallurgy, refractory lining of the outside branches, and better air nozzle design, combined with reducing L/D of the branches piping, have reduced thermal stresses, particularly during start-ups and upset conditions.
- *Cyclones.* Changes in refractory anchor and material, the hanging system, longer L/D, and using more welds in the anchors have improved cyclone performance.

SUMMARY

Improving FCC unit profitability requires operating the unit against as many constraints as possible. Additionally, selective modifications of the unit's components will increase reliability, flexibility, and product selectivity, and will reduce emissions.

REFERENCES

1. Venuto, P. B. and Habib, E. T., Jr., *Fluid Catalytic Cracking with Zeolite Catalysts.* New York: Marcel Dekker, 1979.
2. Lee, S. L., de Wind, M., Desal, P. H., Johnson, C., and Asim, M. Y., "Aromatics Reduction and Cetane Improvement of Diesel Fuels," Akzo Catalyst Chemical Seminar, Dallas, Texas, October 12, 1993.
3. Keyworth, D. A., Reid, T. A., Kreider, K. A., Tatsu, C. A., and Zoller, J. R., "Controlling Benzene Yield from the FCCU," presented at NPRA Annual Meeting, San Antonio, Texas, March 21–23, 1993.
4. Keyworth, D. A., Reid, T., Asim, M., and Gilman, R., "Offsetting the Cost of Lower Sulfur in Gasoline," presented at NPRA Annual Meeting, New Orleans, La., March 22–24, 1992.
5. Reid, T. A., "The Effect of ZSM-5 in FCC Catalyst," presented at World Conference on Refinery Processing and Reformulated Gasolines, San Antonio, Texas, March 23–25, 1993.

6. Engelhard Corporation, "Maximizing Light Cycle Yield," *The Catalyst Report,* TI-814.
7. Engelhard Corporation, "Prediction of FCCU Gasoline Octane and Light Cycle Crude Oil Cetane Index," *The Catalyst Report,* TI-769.
8. Engelhard Corporation, "Controlling Contaminant Sodium Improves FCC Octane and Activity," *The Catalyst Report,* TI-811.
9. Engelhard Corporation, "Catalyst Matrix Properties Can Improve FCC Octane," *The Catalyst Report,* TI-770.
10. Pine, L. A., Maher, P. J., and Wachter, W. A., "Prediction of Cracking Catalyst Behavior by a Zeolite Unit Cell Size," *Journal of Catalysis,* No. 85, 1984, pp. 466–476.

CHAPTER 7

Project Management and Hardware Design Considerations

Since 1942, when the first FCC unit came onstream, numerous process and mechanical changes have been introduced to the unit. These changes were intended to improve the unit's reliability between maintenance turnarounds, to process heavier feedstocks, to operate at higher temperatures, and to shift the conversion to more valuable products. The two critical components of a successful mechanical upgrade or erection of a new unit are effective project management and incorporation of proper design standards.

This chapter addresses project management aspects of a revamp and provides process and mechanical design guidelines that can be used by a refiner in selecting the revamp components of the reactor-regenerator. The objective of a revamp should be a safe, reliable, and profitable operation, although many times the original driving force for a project is to solve a particular mechanical problem or remove a process bottleneck.

7.1 PROJECT MANAGEMENT ASPECTS OF AN FCC REVAMP

The modifications/upgrades to the reactor and regenerator circuit are made for a number of reasons: equipment failure, technology changes, and/or changes in processing conditions. The primary reasons for upgrading the unit are: improving the unit's reliability, increasing the quantity and quality of valuable products, and enhancing operating flexibility.

The revamp or erection of a new unit requires successful execution of the following six phases of the project:

- Preproject
- Process Design
- Detailed Engineering
- Preconstruction
- Construction
- Commissioning/Start-up

7.1.1 Preproject

In the preproject phase, there are a number of actions that a refiner must take "in-house" before embarking upon a mechanical upgrade of an FCC unit. This is particularly true if the revamp's scope contains the use of a new technology. Included in these preproject activities are:

- Identifying the unit's mechanical and process constraints.
- Identifying the unit's operational goals.
- Optimizing the current unit's performance.
- Obtaining a series of validated operating test runs.
- Producing a "statement of requirement" or "revamp objectives" document.
- Selecting an engineering contractor.

In many cases, a refiner decides to revamp a cat cracker and employ a new technology without properly identifying the unit's mechanical and process limitations. Failure to perform a proper constraint analysis of the existing operation can result in focusing on the wrong issues for the revamp. In addition, it is important that the revamp goals match the refinery's overall objectives. The refiner should identify economic opportunities internally before approaching a technology licenser. For example, what is the primary consideration—more conversion, higher unit throughput, or both? At times, a refiner may prefer to satisfy stated objectives internally as opposed to exploring external sources, but all possible sources of a desired product should be explored. It may often be economically more attractive to purchase the desired product from another refiner than to produce it internally. The "market place" can be a less expensive source of incremental supply than the refiner's own in-house production capabilities.

Prior to a mechanical upgrade, the refiner must ensure that, given existing mechanical limitations, the unit's performance has reached its full potential with catalyst and operational changes. It is much easier to determine the effects of the mechanical upgrade with a well-operated unit. Use of more cost-effective changes could achieve the same return as expensive revamp options when an optimized base case is determined.

Any project yield improvements should be based on conducting a series of operating test runs. The test runs should reflect "typical" operating modes. The results should be material/heat balanced. The test runs should be performed shortly prior to the revamp. A comparison of the results, pre- and post-revamp, should reflect no major changes in the catalyst reformulation.

The revamp objectives, constraints, and requirements must be clearly stated in a *statement of requirement* document transmitted to the engineering contractor. The document should be sufficiently detailed and require minimum interpretation so as to avoid oversights and unnecessary site visits.

Selection of a competent engineering contractor to perform process design and detail engineering is a key element in the overall success of a project. Important factors to consider when choosing a qualified contractor are:

- Successful experience in FCC technology and revamps.
- Related experience held by key members of the project team.
- Current and projected workloads.
- Biases and preferences as they relate to proven technologies and suppliers.
- The strength and chemistry of project team members.
- Range of services expected from the contractor, e.g., front-end engineering, detailed engineering, complete E.P.C. through start-up.
- Engineering rate, mark-up, and unit cost of a "change order."

7.1.2 Process Design

Most companies have their own technology for the predesign phase. For the purposes of this book, this phase will be referred to as *front-end engineering design* (FEED). FEED finalizes the process design basis so that the detailed engineering phase can commence. In most cases, FEED is performed by an engineering contractor, but sometimes it is prepared internally by the refiner. The FEED package must be sufficiently completed so that another engineering contractor can finish the detailed engineering with minimum rework.

In a revamp or construction of a new unit which involves a technology upgrade, the engineering contractor commonly supplies a set of product yield projections. Refiners normally use these yield predictions as the basis when conducting economic evaluation and performance guarantee. It is essential that the refiner review these projections carefully to ensure that they agree with the theory and approach expressed by the licenser and that similar yield shifts have been observed by other refiners installing similar technologies. In another words, the refiner should independently check the validity of projected yield improvements.

During the FEED phase of the project, the engineering contractor can be asked to prepare two cost estimates. The initial cost estimate is usually prepared during the very early stages. The accuracy of this estimate is usually plus or minus 40%–50%. This is a factored estimate of equipment and terms of reference. The second cost estimate is prepared at or near the completion of the FEED package. The accuracy of this cost estimate is normally plus or minus 20%. This estimate is usually the basis for obtaining funding for the detailed engineering stage.

The format of the cost estimate is just as important as the content. The format can make a difference when proving whether or not the content is accurate. Therefore, the refiner should require that the contractor present cost estimates in a format that is easy to understand and analyze. In addition, the refiner's cost engineer should independently review the cost estimate to ensure its accuracy and applicability and also to determine the contingency amounts that the owner should maintain in his funding plans.

The FEED package typically consists of the following documents:

- Project scope of work and design basis.
- Process flow diagrams (PFD).
- Feedstock and product rates/properties.
- Utility load data.
- Operating philosophy, start-up, and shutdown procedures.
- List of equipment, materials of construction, and piping classes.
- Piping and instrumentation diagrams (P&ID), tie-in, and line list.

- Instrument index, control valve, and flow element data sheets.
- Electrical load, preliminary instrument, and electrical cable routing.
- Preliminary plot plan and piping planning drawings.
- Specifications and standards.
- Cost estimate.
- Project schedule.

7.1.3 Detailed Engineering

In the detailed engineering stage, the mechanical design of various components is finalized so that the equipment can be procured from the qualified vendors and the field contractor can install it. In preparing construction issue drawings, the designer should pay special attention to avoiding field interferences and allowing sufficient clearance for safety, operability, and maintainability.

To make sure that the project's related safety, health, and environmental issues have been identified and resolved, the refiner should have in effect a process safety program that confirms that the project complies with OSHA requirements.

Procurement of materials in a timely fashion is a necessary part of detailed engineering. Successful procurement requires:

- Early involvement of the procurement team.
- Identification of long-lead and critical items.
- Identification of "approved" vendors.
- Identification of appropriate specification standards.
- Competitive bid evaluation based on quality, availability, and price.
- Establishment of a quality control program to cover fabrication inspection.
- Establishment of an expediting system to avoid unnecessary delays.

7.1.4 Preconstruction

Activities performed in the preconstruction or preturnaround stage are essential to the success of the project. Some of the key activities are:

- Finalizing the project strategy plan.
- Determining required staffing.
- Identifying lay-down needs and securing specific areas.
- Performing the detailed constructibility study.

- Identifying additional resources, such as special equipment or special skills.
- Completing an overall execution schedule.
- Reviewing the schedule to maximize preshutdown work.
- Maximizing preshutdown tasks.

7.1.5 Construction

The guidelines for screening the general mechanical contractor and other associated subcontractors are similar to those for selection of an engineering contractor. The scope and complexity of the work will largely dictate the choice of the general contractor. Aside from availability and quality of skilled crafts, the contractor's safety record and the dedication of the front-line supervisor to the workers' safety should be an important factor in choosing a contractor.

Early selection of the general contractor is critical. The general contractor should be brought in at 30%–40% engineering completion to review the drawings and to interface with the engineering contractor. Additionally, early contractibility meetings among the refiner, engineering contractor, and general mechanical contractor will prove valuable in avoiding delays and reworks.

7.1.6 Precommissioning and Start-up

A successful start-up requires having in place a comprehensive plan that addresses all aspects of commissioning activities. Elements of a such plan include:

- Preparation of the operating manual and procedures to reflect changes associated with the revamp.
- Preparation of training manuals for the operator and support groups.
- Preparation of a field checklist to inspect critical items prior to start-up.
- Development of a QA/QC certification system to assure that the installation has complied with the agreed standards and specifications.

7.1.7 Post-project Review

Shortly after the start-up and before the general contractor leaves the site, a meeting should be held among key members of the project execution team to obtain and document everyone's feedback on what went right, what went wrong, and what could have been done better. A summary of the minutes of this "lessons learned" meeting should be sent to the participants and other relevant personnel.

Once the operation of the unit has "lined out," it is time to conduct a series of test runs to compare performance and economic benefits of the unit with what was projected as part of the original project justification. The results can also be used to determine if the unit's performance meets or exceeds the engineering contractor's performance guarantee.

7.1.8 Useful Tips for a Successful Project Execution

A successful project is defined as one that meets its stated objectives (safety, improved reliability, increased liquid yield, reduced maintenance cost, etc.) on or under budget and is completed on or ahead of schedule. Some of the helpful criteria that ensure a successful project are as follows:

- Plan carefully; this minimizes changes.
- Set the major reviews (PFDs, P&IDs, etc.) early as opposed to waiting until the basic design is completed. This will minimize the project's cost by lessening rework.
- Assign dedicated refinery personnel to be stationed in the engineering contractor's office to coordinate project activities and act as liaison between the refinery and the contractor.
- Make sure the key people from the operations, maintenance, and engineering departments are kept fully informed and that their comments are reflected early enough in the design phase to minimize costly field rework.
- Centralize all decision making to avoid project delays.

7.2 PROCESS AND MECHANICAL DESIGN GUIDELINES

Many aspects of FCC development have been the result of "trial and error." The development of present design standards is as much art as it is science. Consequently, it is appropriate to review some of the key developments that have influenced the current design philosophy behind the FCC reactor and regenerator:

- *Catalyst Quality.* The early FCC catalysts were neither very active nor very selective; yield structure contained too much coke at the expense of gasoline and other valuable products. Regenerators were operating in partial combustion at temperatures of around 1100°F. The introduction of zeolite into the catalyst in the late 1960s brought about a significant impact on the FCC process. The zeolite-based catalyst allowed major yield shifts to light liquid products.
- *Higher-Temperature Operation.* Advances in catalyst technology, the need to process heavier feedstocks, and the need to maximize the yield of desired products have resulted in operating the regenerator and reactor at higher temperatures. These higher temperatures have had deleterious effects on the mechanical components of the reactor/regenerator. The main drawback of a higher temperature operation is a higher induced stress resulting in a lower load-carrying capacity of steel.
- *Refractory Quality.* The refractory lining was first developed mainly for use in the iron and steel industries. It was not until the refractory manufacturers began developing products specifically designed for FCC applications that tremendous improvements in erosion and insulating properties were realized.
- *More Competitive Refining Industry.* The run length of the early FCC units was very short; the unit was shut down every year or so. The general approach at the time was to make the necessary repairs and replace the damaged internal components. Once the industry became more competitive, the drive was to increase the unit's run length, improve its reliability, and maximize the quantity and quality of desired products.

The evolution and improvement of the above-mentioned topics set the background for providing FCC design parameters. The following sections present the latest commercially-proven process and mechanical design recommendations for FCC reactor-regenerator components. These design guidelines, though not universally agreed upon by every FCC "expert," can be useful to the refiner in ensuring that the mechanical upgrade of a unit will be safe, reliable, and profitable.

The components of the reactor-regenerator circuit in which process and mechanical design recommendations are provided are as follows:

- Feed injection system
- Riser and riser termination

- Spent catalyst stripper
- Standpipe system
- Air and spent catalyst distributors
- Reactor and regenerator cyclones
- Expansion joints
- Refractory

7.2.1 Feed Injection System

Any mechanical revamp to improve the unit yields should always begin with installing an efficient feed and catalyst distribution system. This is the single most important component of the FCC unit. An efficient feed and catalyst injection system maximizes gasoline yield and conversion at the expense of lower gas, coke, and decant oil and allows downstream technology to perform at its full potential.

Ideally, a well-designed feed and catalyst injection system will achieve the following objectives:

- Distribute the feed and regenerated catalyst throughout the cross section of the riser to ensure that all feed components are subjected to the same cracking severity.
- Atomize the feed uniformly and instantaneously.
- Avoid recontacting of the "spent catalyst" with the fresh feed.
- Produce proper oil droplet size to penetrate through the catalyst over the 360° cross-sectional area of the riser.
- Avoid erosion of the riser wall and attrition of the catalyst.
- Perform without plugging or erosion.

Process design considerations for feed nozzles

Table 7-1 contains a summary of the process and mechanical design criteria commonly used in specifying high-efficiency feed nozzles. The mechanical design of any feed nozzle should be robust and easy to maintain. Its long-term mechanical reliability is critical in achieving the expected benefits of the upgrade. The following mechanical problems are often encountered: erosion of the nozzle tip(s), erosion of the riser wall, and blockage of the nozzles.

Catalyst lift zone design considerations

To maximize the benefits of feed nozzles, the regenerated catalyst must be distributed evenly throughout the cross section of the riser.

This requires preaccelerating the catalyst to the feed zone. Steam or fuel gas is often used to lift the catalyst to the feed injection. Figure 7-1 shows the design criteria of using steam as a lift media to deliver a "dense" supply of catalyst to the feed nozzles.

7.2.2 Riser and Riser Termination

In most of today's FCC operations, the desired reactions take place in the riser. In recent years, a number of refiners have modified the FCC unit to eliminate or severely reduce post-riser cracking. Quick separation of catalyst from the hydrocarbon vapors at the end of the riser is extremely important in increasing the yield of the desired products. The post-riser reactions produce more gas and coke versus less gasoline and distillate. Presently, there are a number of commercially proven riser disengaging systems offered by the FCC licenser designed to minimize the post-recracking of the hydrocarbon vapors.

The process and mechanical guidelines used in designing most of the new or revamped units are summarized in Table 7-2.

7.2.3 Spent Catalyst Stripper

A properly designed stripper minimizes the quantity of entrained and adsorbed hydrocarbons, which are carried over to the regenerator with the spent catalyst. This goal should be accomplished by the use of

Table 7-1
Process and Mechanical Design Criteria for FCC Feed Nozzles

Injectors	Multinozzles, typically 4 or 6 nozzles per riser located at the periphery of the riser and projected upward.
Pressure drop	40 psi to 60 psi at the design feed rate.
Nozzle exit velocity	150 ft/sec to 300 ft/sec.
Dispersion media and rate	Steam, 1 wt% to 3 wt% of feed rate for convential gas oil. 4 wt% to 7 wt% for residue feedstocks.
Orientation and location	Radial; 3 to 4 riser diameter above the point where the regenerated catalyst enters the riser.
Feed nozzle type	Readily retractable.
Insert material	304 stainless steel.
Nozzle tip	Hard-surfaced inside and out.

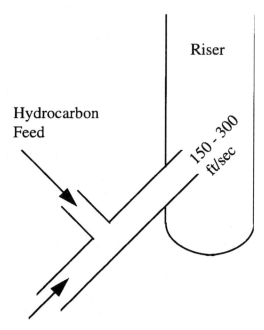

Figure 7-1. Schematic of a typical feed nozzle.

Table 7-2
Process and Mechanical Design Guidelines for FCC Risers

Hydrocarbon Residence Time	1 second to 3 seconds based on the riser outlet conditions. Depending on the degree of catalyst back-mixing in the riser, the catalyst residence time is usually 2.5 to 3.5 times longer than the hydrocarbons.
Vapor Velocity	20 ft/sec minimum (without oil feed), 65 ft/sec to 85 ft/sec at the design feed rate.
Geometry	Vertical—to simulate plug flow and to minimize catalyst back mixing.
Termination	Riser-cyclone separator attached to another separation device to minimize recracking of hydrocarbon vapors.
Configuration	External or internal.
Material	Carbon steel, "cold wall" as opposed to "hot wall."

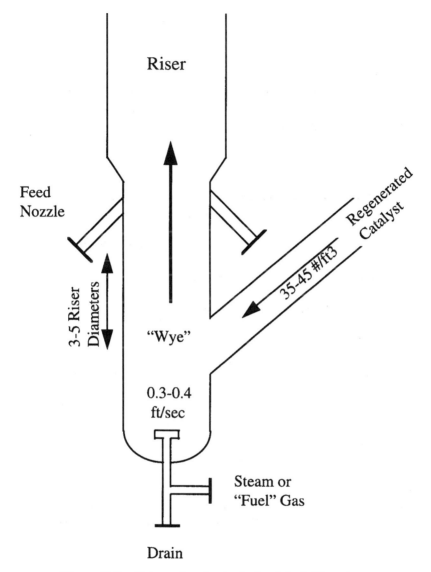

Figure 7-2. Schematic of a typical catalyst lift system.

stripping steam. The major drawbacks of allowing the hydrogen-rich hydrocarbons into the regenerator are losses of liquid products, throughput, and catalyst activity.

Although properly design will greatly enhance stripper performance, it is also very important to note that the performance of the stripper is also largely influenced by type of feedstocks, catalyst, and operating conditions. The key process parameters for designing a stripper are listed in Table 7-3 (see also Fig. 7-3).

Catalyst flux is defined as catalyst circulation rate divided by the "full" cross-sectional area of the stripper. For efficient stripping, it is desirable to minimize the catalyst flux to reduce the carryover of hydrogen-rich hydrocarbons into the regenerator.

Table 7-3
Reactor-Stripper Process and Mechanical Design Criteria

Catalyst Flux	500–700 lbs/min/ft²
Stripping Steam Rate	2–5 lbs of steam per 1,000 lbs of circulating catalyst
Stripping Steam Superficial Velocity	0.5–0.75 ft/sec
Catalyst Residence Time	1–2 minutes
Steam Quality	Superheated ~100°F

Steam Distributor(s)	
Number of Stages	Minimum of two—upper and lower
Type	Pipe grid or concentric rings.
Number of Nozzles	Minimum of one (1) nozzle per ft² of cross-sectional area of the stripper

Nozzles	
Orientation	Two rows, pointing downward
Exit Velocity	125–150 ft/sec
Pressure Drop	Minimum of 1 psi or 30% of the bed height
L/D	Minimum of 5 or long enough to expand "vena contracta"

Material of Construction	
Stripper Shell	Carbon steel, "cold wall" with 4" medium-weight refractory lining.
Distributors	Carbon steel, top distributor externally lined with 3/4" to 1" thick erosion-resistant refractory
Baffles	Carbon steel
Nozzles	Carbon steel, schedule 80 minimum

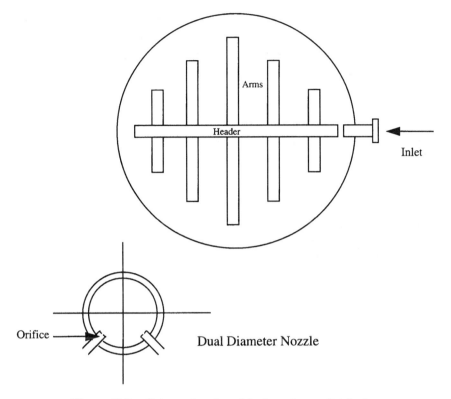

Figure 7-3. Schematic of a stripping steam distributor.

Up to a certain point, stripping efficiency is proportionate to increasing stripping steam rate. However, excess stripping steam overloads the reactor cyclones, main column, and sour water treating system. Therefore, stripping steam should often be varied to determine the optimum rate. The optimum stripping steam rate usually corresponds to a value in which there will be no reduction in the regenerator bed temperature.

The catalyst residence time in the stripper is determined by the catalyst circulation rate and the amount of catalyst in the stripper. This amount usually corresponds to the quantity of the catalyst from the centerline of "normal" bed level to the centerline of the lower steam distributor. A higher catalyst residence time, though it increases hydrothermal deactivation of the catalyst, will improve stripping efficiency.

It is important to note that, depending on the stripper pressure and temperature, a certain fraction of stripping steam is carried with the spent catalyst into the regenerator. Example 7-1 shows how to determine this amount.

Example 7-1

Calculate the amount of entrained stripping steam into the regenerator from a reactor-stripper with the following conditions:

Catalyst skeletal density   = 150 lbs/ft^3
Catalyst flowing density   = 35 lbs/ft^3
Stripper operating pressure   = 25 psig
Stripper operating temperature = 980°F
Catalyst circulation rate   = 40 short tons/min = 4,800,000 lbs/hr

Solution:

Volume of entrained stream = 1/35 − 1/150 = 0.0219 ft^3 of steam/lbs of circulating catalyst

$$\rho = \frac{M}{10.73} \times \frac{P + 14.7}{t + 460}$$

Where:

ρ = Gas or vapor density, lbs/ft^3
M = Molecular weight
P = Pressure, pounds per square inch gauge
t = Temperature, °F

Steam density $= \dfrac{18}{10.73} \times \dfrac{25 + 14.7}{980 + 460} = 0.0462$ lbs of steam$\big/$ft^3 of steam

Entrained steam = (0.0219 ft^3 of steam/lbs of catalyst) × (0.0462 lbs of steam/ft^3 of steam) × 4800000 lbs/hr = 4,858 lbs/hr

---

7.2.4 Standpipe System

Proper standpipe design is one of the most important factors in obtaining good circulation. The standpipe provides the necessary head pressure required to circulate the catalyst. A standpipe assembly is typically comprised of three major parts: hopper, standpipe, and slide or plug valve. The function and design of each part is described below.

Hopper design

The purpose of a catalyst hopper (see Figure 7-4) is to provide sufficient time for the initial deaeration of the catalyst. Proper catalyst deaeration should maximize catalyst density and maintain the catalyst in a "fluidized" state. Table 7-4 contains the key process parameters used in designing standpipe hoppers.

Table 7-4
Process Design Considerations for Standpipe Hoppers

Hopper Entrance Area	4–5 times the standpipe cross-sectional area
Angle of Cone	15°–25° off the vertical
Desired Catalyst Density	40–45 lbs/ft^3
Catalyst Velocity	0.5–1.0 ft/sec

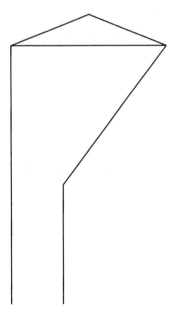

Figure 7-4. Schematic of a typical catalyst hopper.

Standpipe

The standpipe provides the necessary head pressure required to achieve proper catalyst circulation. Standpipes are sized to operate in the fluidized region for wide variation of catalyst flow. Maximum catalyst circulation rates are realized at higher head pressures. The higher head pressures can only be achieved when the catalyst is fluidized. Table 7-5 contains typical process and mechanical design criteria for standpipes.

Table 7-5
Process and Mechanical Design Criteria for Catalyst Standpipes

Catalyst Flux	100–300 lbs/sec/ft²
Catalyst Velocity	2–6 ft/sec (target for 4 ft/sec)
Desired Density	40–45 lbs/ft³
Geometry	Vertical or sloped at maximum angle of 45° (off vertical)
Material	Carbon steel, "cold wall" with 5"-thick heavy-weight, erosion-resistant refractory lining
Supplemental Aeration	Every 5–8 feet along the standpipe, use rotameters to regulate aeration flow

Slide or plug valve

The position of the slide or plug valve regulates the flow of catalyst between reactor and regenerator. The slide valve also provides a positive seal against reversal of the hydrocarbons into the regenerator or hot-flue gas into the reactor. Table 7-6 summarizes typical process and mechanical parameters for designing slide valves.

The formula to calculate catalyst circulation rate through a slide valve is shown below:

$$W = A_p \times C_d \times 2400 \times \sqrt{\Delta P \times \rho}$$

Table 7-6
Process and Mechanical Design Guidelines for Slide Valves

Operating Pressure Drop	Minimum 1.5 psi, maximum 10 psi
% Opening @ Design Circulation	40%–60%
Material	Shell: carbon steel with 4" to 5" thick heavy-weight, single-layer, cast-vibrated refractory with needles.
	Internals: 304H stainless steel for temperature >1200°F and Grade H, 1 1/2% chrome for <1200°F.
	Internal components exposed to catalyst should be refractory-lined for erosion resistance.
	Sliding surfaces should be hard-faced, minimum thickness 1/8".
Bonnet Design	Sloped bonnet (30° minimum) for self draining of catalyst.
Purge	Purgeless design of stuffing box. Guides should be slotted, hard-surfaced, and supplied with purge connections and be normally closed. Nitrogen is the prefered choice of purge gas.
Actuator Type	Electrohydraulic for fast response and accurate control.
Actuator Response Time	A maximum of 3 seconds.

Where:

W = Catalyst circulation rate, lbs/hr
A_p = Port or orifice opening, square inches
Cd = Coefficient of discharge = 0.85
ΔP = Valve pressure drop, psi
ρ = Density of catalyst in the standpipe = lbs/ft^3

Example 7-2

To illustrate the use of the above equation, determine the catalyst circulation rate from the following information:

Slide valve ΔP   = 5 psi

Slide valve opening = 40% corresponding to a port opening of 200 square inches.

Catalyst density   = 35 lbs/ft³

Therefore:

$$W = 200 \times 0.85 \times 2,400 \times \sqrt{35 \times 5} = 5,397,333 \, \text{lbs/hr}$$

$$= 45 \, \text{short tons/min}$$

7.2.5 Air and Spent Catalyst Distributor

The main purpose of the regenerator is to produce a clean catalyst, minimize afterburn, and reduce localized sintering of the catalyst. For efficient catalyst regeneration, it is very important that the air and the spent catalyst are evenly distributed. Although, in recent years, the design of air distributors has improved significantly, the same cannot be said for spent catalyst distributors. This is particularly true in the case of side-by-side FCC units. Most side-by-side units suffer from uneven distribution of the spent catalyst.

A well-designed air distributor system has the following characteristics:

- Distributes the air uniformly to the spent catalyst.
- Withstands mechanically a wide range of operating conditions, including start-up, shutdown, normal operation, and upset conditions.
- Provides reliability with minimum maintenance.

Two types of air distributors have been used in the past. In some of the original side-by-side units, both the air and the catalyst flow through the distributor. In virtually all the air distributors designed today, air only flows through the distributor.

Flat grid plate, dome, pipe, ring, and *flat pipe grid* are the five typical configurations of air distributors being used presently. The most common types are ring and flat pipe grid distributors. Overall, pipe grid is preferred over air ring mainly due to a more uniform coverage and a lower discharge velocity which tends to minimize catalyst attrition. In addition, the pipe grid maintains the same coverage of the cross-sectional area regardless of the air rate. Rings obtain their coverage from jet penetration, and the coverage will be reduced at air rates less than design value due to lower velocity.

The three primary factors affecting the mechanical performance of the air distributor system are: erosion, thermal expansion, and mechanical integrity of supports. The distributor's design should reflect the erosive nature of high catalyst/air velocities, thermal expansion for various operating conditions, and corresponding supports to minimize thermal expansion loads. The process and mechanical design considerations of an air distributor are shown in Table 7-7 (see also Figure 7-5).

7.2.6 Reactor and Regenerator Cyclone Separators

A cyclone separator is an economical device for removing particulate solids from a fluid system. The induced centrifugal force (see Figure 7-6) is tangentially imparted on the wall of the cyclone cylinder. This force increases the density difference between the fluid and solid, thus increasing the relative settling velocity.

Cyclone separators are extremely important to the successful operation of the cat cracker. Their performance can impact several factors, including the additional cost of fresh catalyst, the extra turnaround maintenance costs, allowable limits on emission of the particulate, the incremental energy recovery in the gas compressor, and the hot gas expander.

Designing an "optimum" set of cyclones requires a balance between the desired collection efficiency, pressure drop, space limitations, and installation cost. The cyclone process and mechanical design recommendations are shown in Table 7-8.

7.2.7 Expansion Joint

Efforts should be made to eliminate the use of expansion joints in process piping, however, if needed, the expansion joints are used to mitigate the pipe stresses caused by large thermal movements. Table 7-9 lists the recommended mechanical design criteria for expansion joints.

Table 7-7
Process and Mechanical Design Criteria for Air Distributors

Recommended type	Pipe grid distributor
Nozzle exit velocity	100–200 ft/sec
Pressure drop	1.5–2.0 psi @ design air rate; 10%–30% of the bed static head at minimum air rate for downward-pointing nozzles
Material	304H stainless steel, externally lined with 1"-thick erosion-resistant refractory
Branch pipe	L/D ratio of less than 10 to minimize the support requirement and vibration
Branch arm connection	Continuous pipe through the main header and slotted opening
Fittings	Forged fittings instead of miters for supporting the headers; the forged fittings minimize failures due to stress cracking

Nozzles

Type and orientation	Dual diameter nozzles with orifice in the back of nozzle; downward @ 45°
Length	Minimum of 4 inches
L/D	5/1 to 6/1
Location of first nozzle	8–12 inches from the edge of the slot in the branch arm
Drain size	1/4"–1/2" spaced evenly

The pressure drop of the nozzle's orifice can be calculated from the equation below:

$$\Delta P = \frac{\rho_o}{2 \times g_c \times 144} \times \left(\frac{V_o}{C_d}\right)^2$$

where: V_o = *Velocity of air through the orifice, ft/sec*
ρ_o = *Density of air, lb/ft³*
g_c = *Gravitational constant, 32.2 ft/sec²*
C_d = *Discharge of coefficient = 0.85*

7.2.8 Refractory

In the FCC unit, refractory is used extensively to resist heat, erosion, corrosion, or any combination of the three. The insulating properties protect the reactor-regenerator components from the elevated temperature. The erosion properties protect the equipment from a high-velocity

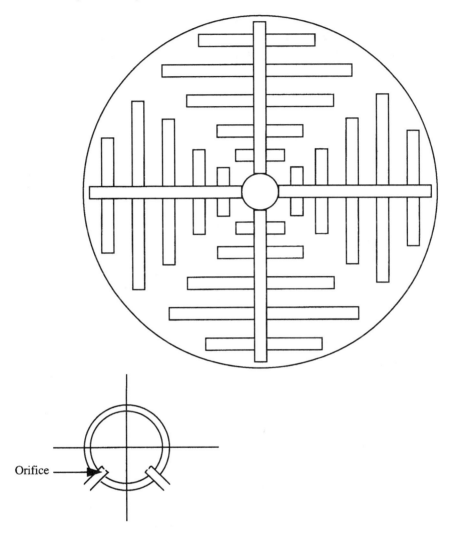

Figure 7-5. Typical layout of a pipe grid distributor.

catalyst. Refractory also provides chemically resistant protection against corrosion reactions.

To ensure a high quality refractory application, the refiner must require the following:

- A mill test report of each pallet of refractory.
- Job-site qualification of all refractory installers before the actual work is performed.

(text continued on page 224)

Table 7-8
Process and Mechanical Design Guidelines
for Reactor and Regenerator Cyclones

Vapor Velocities at Design Feed Rate		
Cyclone Type	Inlet, ft/sec	Outlet, ft/sec
Reactor, single-stage	65–75	100–120
Reactor or regenerator, primary or first-stage	60–70	65–75
Reactor, secondary or second-stage	75–85	100–110
Regenerator, secondary or second-stage	80–85	120–140
Minimum cyclone velocity	25–35	

Minimum Overall Collection Efficiency = 99.995%
Single-stage or primary dipleg mass flux = 100–125 lbs/ft²/sec

Dimensional Specifications			
Parameters	Single-stage	Primary	Secondary
Ratio of cyclone, length/inside diameter	5.0	3.5–4.5	4.5–5.5
Ratio of cyclone inlet, height/inlet width	2.2–2.5	2.2–2.5	2.2–2.5

Material	
Reactor cyclones	Chrome-moly alloy lined with 1"-thick erosion-resistant refractory
Regenerator cyclones	304H stainless steel, lined with 1"-thick erosion-resistant refractory
Reactor Plenum	Chrome-moly alloy, internal or external depending on cyclone geometry.
Regenerator Plenum	Carbon steel, "cold wall" design to avoid high-temperature stress cracking.

Penetration of the gas outlet tube into each cyclone should be at least 80% of the cyclone inlet duct height.
The projected vortex (see Figure 7-6) should be a minimum of 15" above the dust-bowl outlet.
Trickle valves should be partially shrouded.

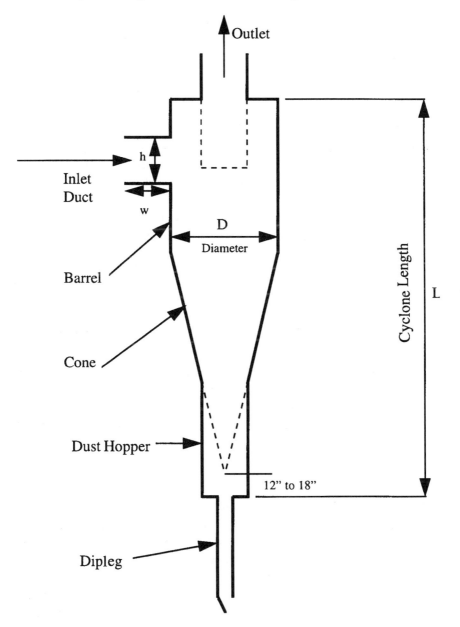

Figure 7-6. Schematic of a typical cyclone.

Table 7-9
Mechanical Design Recommendation for Expansion Joints

Shell's material	Carbon steel, "cold shell design," cast-vibrated 5"-thick refractory lining
Bellow's material	Inconel 625
Purge requirement	Packed bellows, no purge
Configuration of bellows	Two-ply bellows with popout indicator for detecting leakage; each bellows should be capable of maintaining the full pressure
Packing material	Ceramic fiber blanket
Minimum bellows' temperature	400°F to minimize condensation and subsequent acid attack

(text continued from page 221)

- Production test samples to ensure that the required physical properties are met.

To ensure longer service life for the refractory application, a thorough dryout procedure should be implemented and practiced. Slow thermal dryout is required to minimize and eliminate refractory spalling.

General refractory selection guidelines for various components of the reactor-regenerator circuit are listed in Table 7-10.

SUMMARY

Because FCC is one of the most efficient conversion processes, every effort must be made to ensure that its operation is optimized. This may require revamping the existing unit to incorporate the latest proven technologies. Carrying out a successful revamp requires proper identification of the unit's mechanical and process limitations. Failure to perform a proper constraint analysis of the existing operation can result in focusing on the wrong issues for the revamp. The process and mechanical design guidelines detailed in this chapter will be beneficial for the proper selection of the revamp components of the reactor-regenerator.

Table 7-10
Recommended Refractory Specifications for Various FCC Components

Catalyst Carrier Lines (riser, standpipe, wye, slide valves, etc.)	4" to 5" thick, cast-vibrated,single-layer, heavy-weight (140–160 lbs/ft^3),reinforced with stainless steel fibers. The refractory is supported by "wave V footed" type anchors, 304 stainless steel, and spaced on 8" to 12" centers. The anchor's length should 1/2"–1" shorter than the thickness of the refractory lining.
Reactor, Regenerator, and Stripper Shells	4" thick, "gunned," single-layer, medium weight (80–100 lbs/ft^3), with steel fibers, supported by "wave V footed" type anchors with plastic tips.
Reactor and Regenerator Cyclones	Internally lined with 1" thick, erosion-resistant refractory attached by 304 stainless steel hexmetal anchors with every hex on every other row welded independently. The refractory should be cut out flush with the top of hex. The hexmetal should be discernible after the refractory installation.
Cyclone Diplegs	Primary diplegs: internally lined with 1" thick, erosion-resistant lining. Secondary diplegs: lined internally the top 5' from the dust-bowl. Externally, diplegs should be lined where the spent catalyst returns and other known turbulent or wear areas.
Air Distributors	Externally lined with 1" thick, erosion-resistant refractory supported by 304H stainless steel hexmetal anchors or other individually unitized anchors. The refractory should be cut out flush with the top of hex. The hexmetal should be discernible after the refractory installation.
Combustion Air Preheater	4"–5" thick, medium-weight, gunned, with needles, thermal shock-resistant refractory.

REFERENCES

1. Murphy, J. R., Air Products—HRI, "The Development of Feed and Air Distribution Systems in Fluid Catalytic Cracking," presented at the 1984 Akzo Chemicals Symposium, Amsterdam, The Netherlands.
2. Tenney, E. D., General Electric Enviromental Services, Inc., "FCC Cyclone Problems and How They Can Be Overcome with Current Designs," presented at the Grace-Davison FCC Technology Conference, Toledo, Spain, June 3–5 1992.
3. Cabrera, C. A., Hemler C. L., and Davis, S. P., UOP Inc., "Improve Refinery Economics Via Enhanced FCC Operations," presented at Katalistik's 8th Annual FCC Symposium, Budapest, Hungary, June 1–4, 1987.
4. Kalen, B., and O'Broin, A. E., Emtrol Corporation, "A Unique Solution— Cyclone Support in a High-Temperature Environment," presented at Katalistik's 8th Annual FCC Symposium, Budapest, Hungary, June 1–4, 1987.
5. Ratterman, M., "An Approach to the Design and Analysis of Data from the Standpipe System on FCC Units," Gulf Research & Development, Pittsburgh, Pennsylvania, October 1983.
6. Wrench, R. E., and Glasgow, P. E., The M.W. Kellogg Company, "FCC Hardware Options," Paper No. 125C, presented at the AIChE National Meeting, Los Angeles, California, November 17–22, 1991.

CHAPTER 8

Troubleshooting

The cat cracker plays a key role in the overall profitability of the refinery, so it is important that its operation be reliable and efficient. Additionally, the unit must operate safely and in compliance with federal, state, and local requirements regarding emissions and air quality standards. Proper troubleshooting of the cat cracker will help to ensure that this important unit will operate at maximum reliability and efficiency while complying with the environmental concerns.

Troubleshooting deals with identifying and solving immediate problems, as well as recognizing ways to optimize daily performance of the unit. Problems can be related to management, operation, hardware and equipment, or process issues. Solutions to these problems can take many forms, such as improved operating procedures and training, implementation of a preventative maintenance program, or installation of state-of-the-art equipment to improve unit performance and/or avoid a shutdown.

The objective of this chapter is to outline the fundamental steps toward effective troubleshooting and to provide the reader with a practical and systematic way to approach a problem and develop a solution. General guidelines are provided for identifying most problems and determining a diagnosis. In particular, the following problem areas are addressed in this chapter:

- Catalyst Circulation
- Catalyst Loss
- Coking/Fouling
- Flow Reversal
- High Regenerator Temperature
- Afterburn
- Hydrogen Blistering
- Hot Gas Expander
- Amount and Quality of Products

8.1 GUIDELINES FOR EFFECTIVE TROUBLESHOOTING

A successful troubleshooting assignment will require someone to:

- Be a good listener.
- Gather historical background.
- Evaluate "common" and "uncommon" causes of problems.
- Examine goals and constraints to verify the applicability of the present operation.

The management, engineering, and operations departments may perceive a problem differently. Frequently, there is someone familiar with the operation who most likely knows the symptoms and possibly can offer a solution to the problem, but for various reasons, people who are in a position to implement the solution may not have thought to ask him or her. Typically, those closest to the problem are the unit operator and the maintenance foreman; they could offer valuable input. Do not draw any conclusions before gathering all applicable facts.

Examine similar problems that have previously occurred in the system to determine how they were diagnosed and solved. Review the operating and maintenance records and compare performance of the normally operating unit to the current problematic operation. Using reliable historical data as a tool helps to identify and diagnose the problem.

Begin by listing all potential causes, using a brainstorming approach. Then, systematically rule out each cause. Brainstorming will help to identify potential causes that have not been previously considered. Be sure not to eliminate uncommon causes without further investigation. Additionally, ensure that limits outlined by process and equipment documentation are consistent with the actual operation of the unit.

The problems often encountered in a cat cracker are due mainly to changes in the *feedstock, catalyst, operating variables,* and *mechanical equipment.* As stated previously, the solution can take the form of improving yields, avoiding shutdowns or, increasing unit reliability.

8.2 CATALYST CIRCULATION

Catalyst circulation to a cat cracker is like blood circulation to a human body. Troubleshooting circulation problems requires a good understanding of the pressure balance around the reactor-regenerator circuit and the factors affecting catalyst fluidization. The fundamentals of fluidization and catalyst circulation are discussed in Chapter 5.

Catalyst circulation is broadly influenced by the physical layout of the unit and the fluidization properties of the catalyst. Depending on the configuration, some cat crackers circulate with ease regardless of the catalyst's physical properties. However, in other designs, the unit can experience circulation difficulties with minor changes in catalyst properties.

There are two types of circulation problems in the FCC unit that require troubleshooting. The first occurs when the apparent maximum circulation has been reached, and the second occurs when the unit is experiencing erratic circulation.

8.2.1 Circulation Limitations

Often, the need to increase the feed rate or the conversion cannot be achieved because the unit is circulation-limited. Effective troubleshooting of the circulation limitation requires having in place a methodology such as the one shown in Figures 8-1A and 8-1B.

To begin the troubleshooting, one must first start by conducting a single-gauge pressure survey of the reactor-regenerator and the main column circuits as discussed in Chapter 5. The results of the survey should provide the foundation for pinpointing the cause(s) and identifying solutions.

The demand for higher catalyst circulation usually requires a higher opening of the regenerated and spent catalyst slide (or plug) valves. The higher circulation increases the pressure drop in the riser and in the reactor cyclones, thus lowering the differential pressure across the slide valves. To ensure proper control and to protect from the reversal of hydrocarbons to the regenerator or the back-flow of hot flue gas to the reactor, slide valves are typically operated within a 25% to 60% opening and a pressure differential of 3 to 6 psi. Any increase in the throughput or the conversion will either increase the opening, reduce the differential, or both.

Causes of insufficient circulation

A lower pressure differential across the slide valve and/or a higher than "normal" slide valve opening are commonly evidence of a unit's

(text continued on page 232)

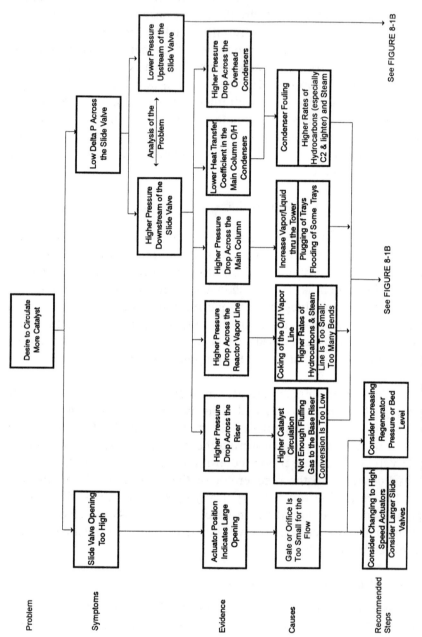

Figure 8-1A.   Troubleshooting circulation limitations.

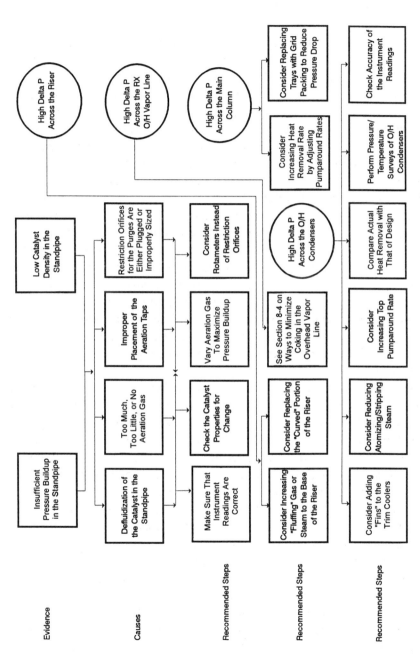

Figure 8-1B. Troubleshooting circulation limitations.

(text continued from page 229)

reaching its circulation limitation. The causes of low differential are either an insufficient pressure buildup upstream of the slide valve or too high a pressure drop downstream of the slide valve.

Insufficient pressure generation upstream of the slide valve is largely due to too little, too much, or no aeration gas in the standpipe. In a properly designed standpipe, the flow of the catalyst develops a smooth and uniform static head over the entire length of the standpipe. This buildup of the pressure head provides the necessary driving force for catalyst circulation. However, as the catalyst travels down the standpipe, it can lose some of its fluidity due to the compression of interstitial gas being carried with the catalyst. This is particularly true in a long standpipe and in regenerators operating at low pressure.

To retain fluidity of the catalyst and to maintain catalyst densities in the 35 to 45 lbs/ft^3 range, many standpipes require external aeration gas to be injected into the down-flowing catalyst. The quantity of aeration medium and correct location of aeration taps are essential in achieving optimum catalyst density (see Figure 8-2). External aeration is not ordinarily needed in short standpipes (less than 30 feet) because, usually, sufficient gas is drawn into the standpipe to keep the catalyst fully fluidized.

Restriction orifices are frequently employed to distribute aeration gas into the standpipes. The orifices are sized for critical flow so that a constant flow of aeration gas is injected regardless of changes in the downstream pressure. Nonetheless, restriction orifices could be oversized, undersized, or plugged with catalyst resulting in overaeration, underaeration, or no aeration. All these phenomena cause low-pressure buildup and low slide valve differential.

Sometimes, a lack of sufficient differential across the regenerated catalyst slide valve is not due to inadequate pressure buildup upstream of the valve but rather an increase in pressure downstream of the slide valve. Possible causes of this increased back pressure are an excessive pressure drop in the riser, reactor overhead vapor line, main fractionator, and/or the main fractionator overhead condensers.

The riser pressure drop is related mainly to the catalyst circulation rate and the *slip factor.* Slip factor is defined as the ratio of catalyst residence time in the riser to the hydrocarbon vapor residence time. Some of the factors affecting the slip factor are circulation rate, riser

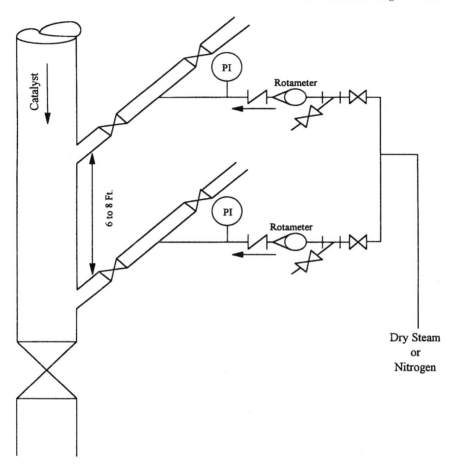

Figure 8-2. Typical standpipe aeration.

diameter/geometry, and riser velocity. Higher than normal pressure drop in the riser could also be due to insufficient fluidization gas in the base of the riser. This is particularly true in the units with radial-design feed nozzles, where the amount of fluffing gas will vary the catalyst density below the feed nozzles and thus the pressure buildup downstream of the slide valve.

The pressure drop across the cyclones, reactor vapor line, main fractionator, and main column overhead condensers is primarily a function of vapor velocity and any partial plugging caused by coking and/or fouling.

Troubleshooting steps

An effective troubleshooting of the circulation limitation requires following a methodology similar to the procedures outlined in Figures 8-1A and 8-1B. Some of the key steps are as follows:

- Obtain a pressure/density profile upstream and downstream of the slide valves.
- Verify any changes in catalyst physical properties.
- Ensure that the correct amount of aeration gas is injected along the standpipes. This is done by varying the aeration flow until maximum slide valve differential is observed.

8.2.2 Erratic Circulation

Erratic circulation occurs when the catalyst is not developing a smooth and uniform static head over the entire length of the standpipe. When this happens, the catalyst packs and bridges across the standpipe. Symptoms of erratic circulation are as follows:

- Severe vibration and movement of the standpipes.
- A noise similar to train chugging.
- Sudden loss of the pressure above the slide valve.
- Fluctuation in the slide valve delta P.
- Ragged reactor temperature and/or stripper level control.
- Pressure swings in the regenerator and the gas plant.

Several factors contribute to erratic circulation. Included are:

- A foreign object such as a piece of refractory has partially obstructed the flow of catalyst in the standpipe.
- Improper aeration—either too much or too little.
- The catalyst has become too coarse. The coarseness or the lack of fines could be mainly due to changes in the catalyst's physical properties and/or malfunctioning of the cyclones.

To troubleshoot erratic circulation, one must:

- Make sure that neither too much nor too little aeration gas is being applied.
- Verify that the fresh catalyst properties have not changed.
- Verify any recent design changes in the standpipes and/or catalyst hopper.
- Check recycling of the regenerator fines and/or the slurry recycle.

Figure 8-3 shows a step-by-step approach to troubleshooting erratic circulation.

8.3 CATALYST LOSSES

Catalyst losses from the FCC unit will have adverse effects on the unit operation, environment, and catalyst cost. Catalyst losses appears as excessive carryover to the main fractionator or losses from the regenerator. Evidences of catalyst losses are:

- An increase in the ash and BS&W content of the slurry oil.
- An increase in the recovery of catalyst fines from the electrostatic precipitator or the tertiary separator.
- An increase in the opacity of the precipitator stack gases.
- A decrease in the 0 to 40 microns fraction of the equilibrium catalyst or an increase in average particle size.
- A gradual loss of the catalyst level in the reactor stripper and/or in the regenerator.

8.3.1 Causes of Catalyst Losses

The common causes of the catalyst losses are due to the following:

- Changes in catalyst properties.
- Changes in operating conditions.
- Changes in mechanical conditions of the unit.
- Changes in operating practice.

Changes in the fresh catalyst's physical properties may contribute to catalyst losses. The losses could be due to the fresh catalyst's being "soft." "Softness," evidenced by the quality of the catalyst binder and the large amount of 0–40 microns, will increase the attrition tendency of the catalyst and thus its losses.

Changes in operating parameters also affect catalyst losses. Examples are:

- An increase and/or decrease in catalyst loading to the cyclones. Overloading the cyclones, even at a constant/or higher efficiency, will result in higher catalyst losses.
- An increase in the feed atomizing steam, causing catalyst attrition and generating fines.
- Addition of a large amount of steam to the regenerator, again causing catalyst attrition.

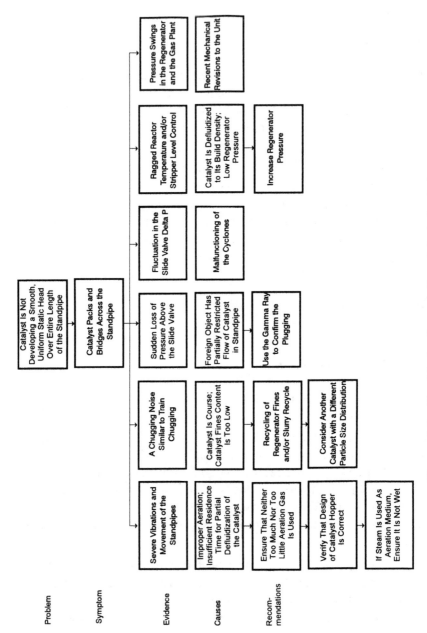

Figure 8-3. Troubleshooting erratic catalyst circulation.

Often, the main causes of catalyst losses are due to changes in the *mechanical conditions* of the unit, such as the cyclones, air/steam distributor, and feed nozzles. Examples are:

- Trickle valves are stuck either "closed" or "open," possibly due to the hinge's being warped or bound.
- Trickle valves are either warped or eroded. Warpage could be due to exposure to high temperature. Erosion could be due to excessive gas leakage into the diplegs.
- Trickle valves have fallen off due to inadequate bracing and/or high superficial gas velocity in the regenerator.
- Holes have formed in the diplegs because of high cyclone velocity and/or mass flow.
- Spalled coke or refractory is lodged in the diplegs due to the introduction of the feed to the unit too early or to inadequate refractory supports.
- Cracks have formed in the internal plenum, possibly due to the load induced by thermal stresses.
- Dipleg diameter is either too small or too large.
- Improperly designed, eroded, or even missing restriction orifices used for steam purges could cause catalyst attrition. Catalyst attrition is also caused by a broken air distributor and high catalyst velocities through slide valves.

8.3.2 Troubleshooting Catalyst Losses

To stop excessive catalyst losses, one must identify whether the loss is from the reactor or the regenerator. In either case, the following general guidelines should be helpful in troubleshooting catalyst losses:

- Plot the physical properties of the equilibrium catalyst. The plotted properties will include particle size distribution and apparent bulk density. The graph confirms any changes in catalyst properties.
- Have the lab analyze the "lost" catalyst for particle size distribution. The analysis will provide clues as to the sources and causes of the losses.
- Check the cyclone loading with that of the design. If the vapor velocity into the reactor cyclones is low, consider adding supplemental steam to the riser. If the mass flow rate is high, consider

taking steps such as increasing the feed preheat temperature to reduce catalyst circulation.

- Confirm that the restriction orifices used for instrument purges are in proper working condition and that the orifices are not missing.
- Verify the accuracy of the catalyst levels in the regenerator and in the reactor-stripper.
- Consider switching to a harder catalyst. For a short-term solution if the losses are from the reactor side, consider recycling slurry to the riser. If the catalyst losses are from the regenerator, consider recycling catalyst fines to the unit.

Figure 8-4 is a summary of the above discussions.

8.4 COKING/FOULING

Nearly every cat cracker at one time or another has experienced some degree of coking/fouling. The common places in which coke has been found are in the reactor walls, dome, cyclones, overhead vapor line, and the slurry bottoms pumparound circuit. Depending on the severity, coking/fouling can have an adverse impact on the refinery's operating profit and maintenance costs, due to cutback in throughput and extra downtime.

8.4.1 Evidence of Coking/Fouling

There are several ways in which coking/fouling in the reactor and the main column can be detected. Examples are as follows:

- Cavitation and/or loss of the main column bottoms pumps.
- Fouling and subsequent loss of heat transfer coefficient in the bottoms pumparound exchanger.
- High pressure drop across the reactor overhead vapor line.
- Excessive catalyst carryover to the main column.

8.4.2 Reasons for Coking/Fouling

The main reasons for the formation of coke in the reactor and in the main column bottoms pumparound are:

- Changes in operating parameters.
- Changes in catalyst properties.

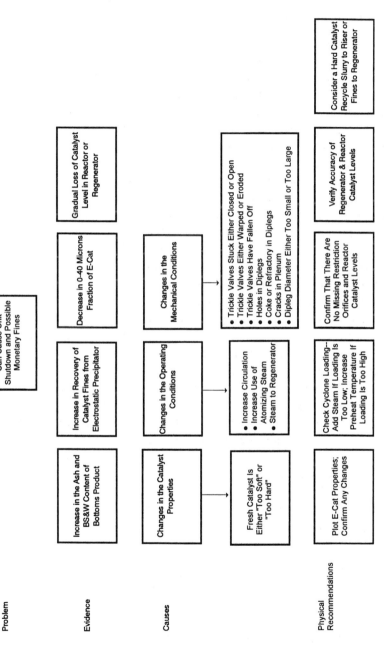

Figure 8-4. Troubleshooting catalyst losses.

- Changes in feedstock properties.
- Changes in mechanical conditions of the equipment.

The operating conditions of the unit, particularly during startups and feed interruptions, will have a large influence on the formation of coke. Coke normally grows wherever there is a cold spot in the reactor system. When the temperature of the metal surfaces in the reactor walls and/or the vapor line falls below the dew point of the vapors, condensation of the reaction products occurs. Condensation and subsequent coke buildup are due to cooling effects at the surface.

A high fractionator bottoms level, a low riser temperature, and a high residence time in the reactor dome/vapor line are additional operating factors that increase coke buildup. If the main column level rises above the vapor line inlet nozzle, a "donut" shaped coke can form at the nozzle entrance to the main column. A low reactor temperature may not fully vaporize the feed; unvaporized feed droplets will aggregate to form coke on the reactor walls and/or the transfer line. A long residence in the reactor and transfer line also accelerates coke buildup.

Certain catalyst properties appear to increase coke formation. Catalysts with high rare earth content tend to promote hydrogen transfer reactions. Hydrogen transfer reactions are bimolecular reactions which can produce multi-ring aromatics.

The quality of the FCC feed also impacts coke buildup in the reactor internals and vapor line and fouling/coking of the main column circuit. The asphaltene or the resid content of the feed, if not converted in the riser, could contribute to this coking. Damaged or partially plugged feed nozzles can contribute to coke formation due to poor feed atomization.

8.4.3 Troubleshooting Steps

The following are some of the steps that can be taken to minimize coking/fouling:

- *Avoid dead spots.* Coke grows wherever there is a cold spot in the system. Use "dry" dome steam to purge hydrocarbons from the stagnant area above the cyclones. Dead spots cause thermal cracking.
- *Minimize heat losses from the reactor plenum and the transfer line.* Heat loss will cause condensation of heavy end components of the reaction products; make sure inlet nozzle to the main column is well insulated.

- *Improve the feed/catalyst mixing system and maintain a high conversion.* A properly designed feed/catalyst injection system, combined with operating at a high conversion, will crack out high-boiling feeds that otherwise could be the precursors for formation of coke.
- *Follow proper start-up procedures.* Introduce feed to the riser only when the reactor system is adequately heated up. Local cold spots cause coke to build up in the reactor cyclones, the plenum chamber, or the vapor line.
- *Keep the tube velocity in the bottoms pumparound exchanger(s) greater than 7 ft/sec.*
- *Hold the main column shed tray's liquid temperature under 700°F.* Ensure adequate wash to shed decks to minimize coking in the bottom of the main column.
- *Utilize a continuous-cycle oil flush into the inlet of the bottoms exchanger.* This keeps the asphaltenes in solution and increases tube velocity.

Figure 8-5 is a summary of the above discussions.

8.5 FLOW REVERSAL

It is essential to maintain a stable pressure differential across the spent and regenerated catalyst slide valves. The direction of catalyst flow must always be from the regenerator to the reactor and from the reactor-stripper back to the regenerator. A negative differential pressure across the regenerated catalyst slide valve would allow fresh feed and oil-soaked catalyst to backflow from the riser into the regenerator. This flow reversal can result in uncontrolled burning in the regenerator and potentially damage regenerator internals, costing a refiner several million dollars in production loss and maintenance expense.

Similarly, a negative pressure differential across the spent catalyst slide valve would allow hot-flue gas backflow to the reactor and the main fractionator, severely damaging the mechanical integrity of these vessels. Some of the main causes of loss of pressure differential across the slide valves are as follows:

- Loss of an air blower.
- Presence of water in the feed.
- High catalyst circulation rates which result in excessive slide valve opening and low differential.

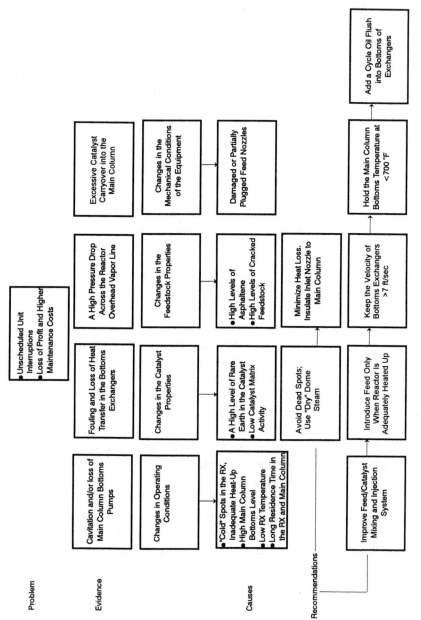

Figure 8-5. Troubleshooting coking/fouling.

- Loss of regenerator or stripper bed levels.
- Failure of the reactor temperature controller and reactor-stripper level controller.

Troubleshooting flow reversal is outlined in Figure 8-6.

8.5.1 Reversal Protection Philosophy

Due to the complex nature of the FCC process, there are many scenarios which can result in upset operations. If the upset condition is not corrected or controlled, each scenario could lead to a reversal. Table 8-1 contains a cause/effect shutdown matrix indicating scenarios in which a shutdown (reversal) could take place. In most cases, a unit shutdown is not necessary, provided there is adequate warning (alarms) of out-of-range critical control elements and that operating staff is competent and sufficiently trained to effectively respond to those warnings.

Slide valves will have an independent low differential pressure override controller to prevent the reaction temperature controllers from opening the slide valves to the point where low differential pressure could allow feed back to the regenerator.

8.6 HIGH REGENERATOR TEMPERATURE

The regenerator "dense phase" temperature (see Figure 8-7) can be adversely affected by the feed quality, the condition of the catalyst, the operating variables, and the mechanical conditions of the unit:

• Feedstock	A higher fraction of $1050°F^+$ in the feed.
• Catalyst	A higher level of rare earth or an increase in the matrix content.
• Operating Variables	Reduced stripping steam or atomizing steam flow. A higher preheat or riser temperature.
• Mechanical Conditions	Damaged stripping steam distributor and/ or damaged feed nozzles.

(text continued on page 247)

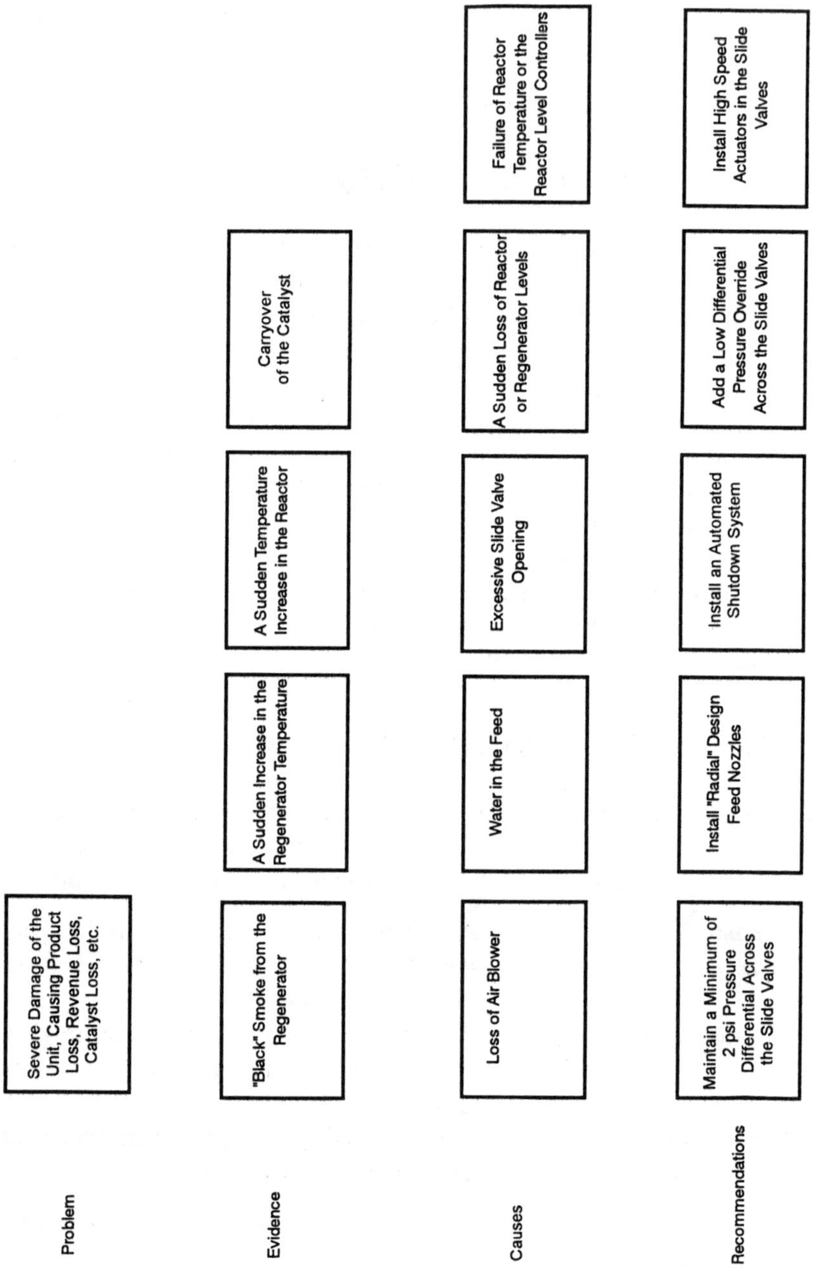

Problem	Severe Damage of the Unit, Causing Product Loss, Revenue Loss, Catalyst Loss, etc.		
Evidence	"Black" Smoke from the Regenerator	A Sudden Increase in the Regenerator Temperature	A Sudden Temperature Increase in the Reactor	Carryover of the Catalyst
Causes	Loss of Air Blower	Water in the Feed	Excessive Slide Valve Opening	A Sudden Loss of Reactor or Regenerator Levels	Failure of Reactor Temperature or the Reactor Level Controllers
Recommendations	Maintain a Minimum of 2 psi Pressure Differential Across the Slide Valves	Install "Radial" Design Feed Nozzles	Install an Automated Shutdown System	Add a Low Differential Pressure Override Across the Slide Valves	Install High Speed Actuators in the Slide Valves

Figure 8-6. Troubleshooting flow reversal.

Table 8-1
A Cause and Effect Shutdown Matrix

Cause \ Effect	Close Riser Regenerated Catalyst Slide Valve	Open Riser Emergency Steam Valve	Close Feed to Riser	Close Slurry Recycle Valve	Close HCO Recycle Valve	Close Spent Catalyst Slide Valve	Open Regenerator Emergency Steam Valve	Alarm Only
Regenerated Catalyst Slide Valve Low Differential Pressure								X
Spent Catalyst Slide Valve Low Differential Pressure								X
Air Blower Low/Low Air Flow	X	X	X	X	X	X	X	
Riser Low/Low Feed Flow	X	X	X	X	X	X	X	
Low Reactor Temperature								X
Reactor Vessel High Catalyst Level								X
Manual Shutdown	X	X	X	X	X	X	X	

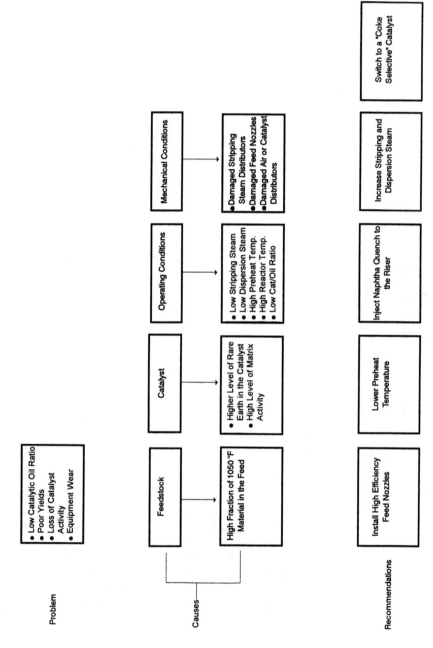

Figure 8-7. Troubleshooting high regenerator temperature.

(text continued from page 243)

8.7 INCREASE IN AFTERBURN

The degree of afterburn in the regenerator largely depends on the operating conditions of the unit and the effectiveness of the contact between the combustion air and the spent catalyst.

- Operating Variables   Excess oxygen, particularly in partial combustion mode, low catalyst residence time in the regenerator, low regenerator bed temperature, and insufficient promoter.
- Mechanical Conditions   Damaged or improperly designed air and/or spent catalyst distributors.
- Troubleshooting Steps   Increase regenerator catalyst bed level, increase the preheat temperature, increase regenerator pressure, increase stripping steam rate, add slurry recycle (see Figure 8-8).

8.8 HYDROGEN BLISTERING IN THE GAS PLANT

Hydrogen blistering is the result of cyanide-induced corrosion in the vapor recovery unit. The simplified chemical reactions are as follows:

$$Fe + 2HS^- \rightarrow FeS + S^{-2} + 2H$$

$$FeS + 6CN^- \rightarrow Fe(CN)_6^{-4} + S^{-2}$$

Normally, a layer of FeS scale, produced in the first reaction above, would protect the interior of the pipe or vessel from further reacting with H_2S. However, the cyanide reacts with FeS to remove this protective scale, thus exposing more free iron to react with H_2S. This cycle continues until blistering, cracking, and eventual total corrosion of equipment occurs.

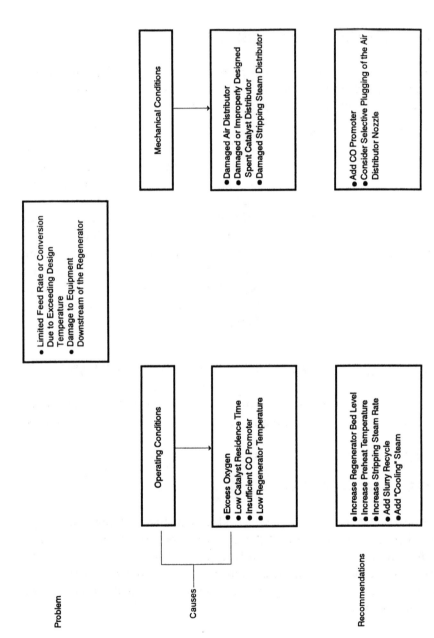

Figure 8-8.   Troubleshooting excessive afterburn.

8.8.1 Causes of Higher Levels of Cyanide

The primary sources of higher cyanide production are:

• Higher levels of nitrogen in the feed.
• Higher reactor temperature.
• Operating in partial combustion.
• Higher matrix activity of the catalyst.

8.8.2 Steps to Control Hydrogen Blistering

The best way to minimize hydrogen blistering is to control the corrosion rate. Both corrosion and hydrogen blistering rates can be reduced significantly by implementing the following steps (see also Figure 8-9):

1. Install a properly designed water-wash system for dilution and removal of cyanide from the unit. Do not use wash water from the high-pressure zone and recycle it back either into the main column overhead or to the outlet first-stage wet gas compressor. Use fresh water and pump it from low pressure to high pressure.
2. Add polysulfide solution to neutralize the cyanide.
3. Install and monitor corrosion and hydrogen probes in the key areas.
4. Inject a filming amide to provide a protective barrier that will prevent cyanide from contacting the iron sulfide layer.

8.9 HOT GAS EXPANDERS

In a number of cat crackers, power recovery trains have been installed to recover energy from the flue gas. However, a significant deposition of catalyst fines in the expander will lead to serious blade wear, power loss, and rotor vibration. Deposit occurs mostly where flue gas velocities are at maximum levels, such as blade outer diameter.

8.9.1 Causes of Blade Wear, Power Loss, and Rotor Vibration

1. Increase in catalyst loading to the regenerator cyclones and third-stage separator.
2. Increase in flue gas rate.

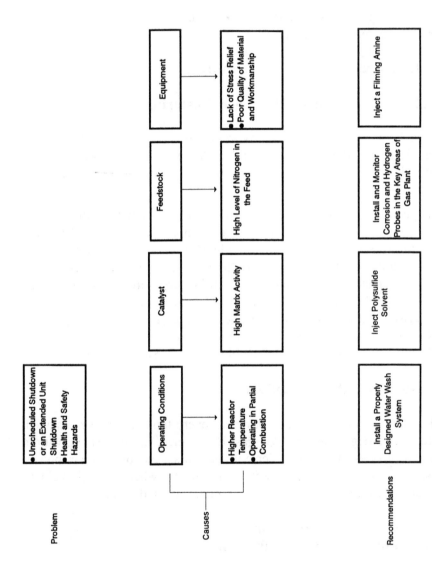

Figure 8-9.   Troubleshooting hydrogen blistering.

3. Increase in fresh catalyst addition rate.
4. A too "soft" or too "hard" catalyst.
5. Sodium and vanadium in the catalyst, resulting in formation of low-melting eutectic which makes the catalyst very sticky.

8.9.2 Troubleshooting Steps

1. Regular monitoring of rotor blade conditions by visual inspection, taking photographs, and/or video recording.
2. Periodic monitoring of rotor casing vibration, bearing temperature, and the expander inlet/outlet temperatures.
3. Periodic monitoring of third-stage separator performance. Consider using more than the "standard" 3% flue gas underflow.
4. On-line cleaning—injection of walnut hulls into the inlet of the expander weekly.
5. Thermal shocking—reducing feed in an approximate 20% increment while maintaining maximum air rate to the regenerator. Cooling at a rate that maintains expander inlet temperature to around 1000°F and holding it for at least one hour.

Figure 8-10 provides an outline of the above discussion.

8.10 PRODUCT QUANTITY AND QUALITY

The amount and quality of products obtained from the FCC unit are influenced largely by feed quality, catalyst properties, operating variables, and mechanical conditions of the unit. The indicators which are often employed to measure the unit's performance are:

• Conversion
• Dry gas yield
• Gasoline yield
• Gasoline octane

8.10.1 Observing a Low Conversion

The "true" conversion is affected by feed quality, catalyst, operating variables and mechanical conditions (Figure 8-11A).

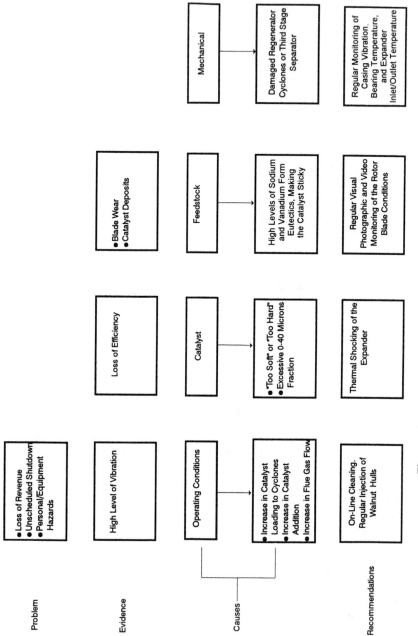

Figure 8-10. Troubleshooting the hot gas expander.

Feedstock quality

The feed properties that most adversely affect conversion are:

- Increase in residue (1050°F+) content.
- Increase in feed impurities such as nickel, vanadium, sodium, or nitrogen.
- Increase in naphthene and aromatic fractions of the feed.

Catalyst properties

Both the fresh and E-cat catalyst properties have a large influence on the unit's conversion. The prime properties are:

- Decrease in microactivity.
- Decrease in the surface area.
- Increase in the CRC.

The decreases in microactivity and surface area are strong functions of thermal deactivation occurring in the regenerator and in the presence of metals in the feed.

Operating variables

The following operating parameters contribute to a decrease in conversion:

- Decrease in the reactor temperature.
- Decrease in the catalyst-to-oil ratio.
- Decrease in the atomizing steam.
- Decrease in the fresh catalyst addition rate.

Mechanical conditions

A damaged or plugged feed nozzle(s) and/or damaged stripping steam distributor(s) are the common causes of mechanical failures that affect the "true" conversion. Note that the "apparent" conversion, as discussed in Chapter 5, is affected by the distillation cut point and the main column operations.

Troubleshooting steps

- Trend the feedstock properties; look for changes in the K factor, 1050°F+, aniline point, refractive index, and °API gravity.

- Plot properties of the fresh and equilibrium catalysts; ensure that the catalyst supplier is meeting the agreed quality control specifications. Verify the fresh makeup rate. Check recent temperature excursions in the regenerator.
- Trend the reactor temperature, cat-to-oil ratio, and atomizing steam rate. Verify the accuracy of the reactor temperature thermocouple and atomizing steam flow meter.
- Perform a single-gauge pressure survey around the feed nozzles. Calculate the hydrogen content of the spent catalyst. Conduct a gamma ray scan test to verify the mechanical condition of the stripping steam distributor.

8.10.2 Observing a High Dry Gas Yield

Like conversion, the dry gas yield is also affected by feed quality, catalyst properties, operating variables, and mechanical conditions (Figure 8-11B).

Feedstock quality

The feed parameters which increase the dry gas yield are:

- Increase in nickel and vanadium content.
- Increase in naphtene, olefin, and aromatic concentration. This is indicated by an increase in refractive index and decreases in aniline point and K factor.

Catalyst properties

The E-cat properties having the greatest negative influence on the dry gas yield are:

- Increase in the level of nickel, vanadium, and sodium.
- Decrease in E-cat activity, surface area, fresh catalyst activity, and rare earth content.
- Increase in the gas and coke factors of the E-cat.

Operating variables

The operating parameters having the most negative impact on dry gas yield are:

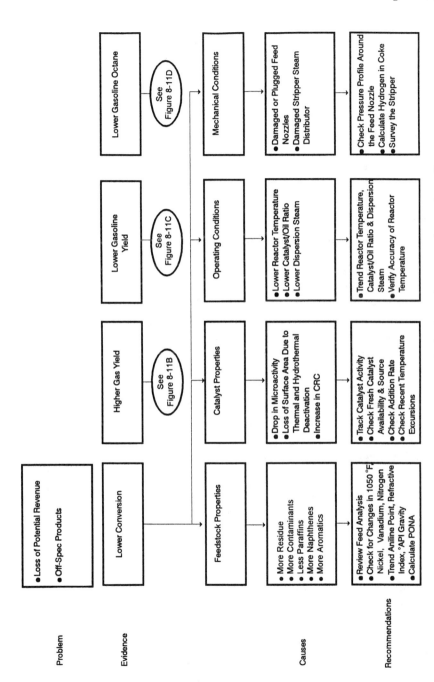

Figure 8-11A. Troubleshooting desired product quantity and quality.

- Increase in the reactor temperature.
- Increase in the regenerator temperature.
- Decrease in the atomizing steam flow.
- Increase in slurry or HCO recycle.

Mechanical conditions

Adverse effects on dry gas yield can be due to:

- A failing reactor temperature thermocouple.
- Partially plugged or damaged feed nozzles.

Troubleshooting steps

The following steps should be carried out:

- Track changes in the metals content of the feed and trend the feed's aniline point and refractive index.
- Trend changes in the catalyst activity, surface area, rare earth, and metals content.
- Verify the position of the wet gas compressor kick-backs. Determine if the compressor turbine needs water-washing. Verify the shift in the amount of inert gases from the reactor.
- Have the instrument technician calibrate the reactor temperature readings. Conduct a pressure survey around the feed nozzle piping to verify its mechanical integrity.

8.10.3 Observing a Lower Gasoline Yield

The FCC "true" gasoline yield also depends largely on changes in feed quality, catalyst properties, operating variables, and mechanical conditions (Figure 8-11C).

Feedstock quality

Paraffinic feedstocks produce the most gasoline yield. The common indicators of an increase in feed paraffinicity are:

- Increase in the K factor.
- Increase in the aniline point.
- Increase in the nickel-to-vanadium ratio.
- Decrease in the fraction of "cracked" material.

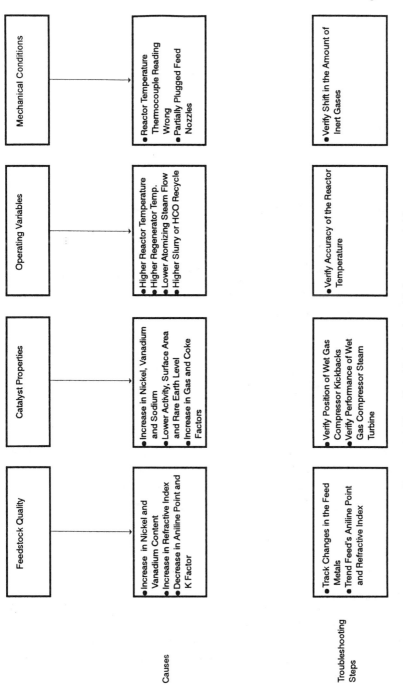

Figure 8-11B. Troubleshooting higher gas yield.

Catalyst properties

The fresh catalyst properties that cause an increase in gasoline yield are:

- Increase in the rare earth content.
- Increase in the zeolite content.
- Increase in the unit cell size.

Operating conditions

The main operating parameters that favor gasoline yield are:

- Decrease in the feed preheat temperature and subsequent increase in the catalyst-to-oil ratio.
- Decrease in the carbon content of the E-cat if the carbon is greater than 0.1 wt%.
- Increase in the reactor temperature if overcracking is not occurring.
- Decrease in the ZSM-5 additive—a shift in FCC gasoline at the expense of LPG.

Mechanical conditions

Deterioration of the feed nozzles or erroneous reactor catalyst level will greatly reduce the gasoline yield.

Troubleshooting steps

- Trend the feed °API gravity, K factor, and aniline point. Verify any changes in paraffin content of the feed.
- Plot the catalyst's unit cell size, rare earth, and activity. Check if there is any fluctuation in catalyst properties.
- Verify the gasoline end point, vapor pressure, and distillation of the LCO to ensure minimum undercutting of gasoline in the LCO.

8.10.4 Observing a Low Gasoline Octane

In general, any parameter which increases the gasoline yield will also decreases its octane; one reason is that the high-octane components in the gasoline tend to be denser than the low-octane components. Therefore, any change which produces more gasoline will result in a lower octane. Again, feedstock, catalyst, operating variables, and

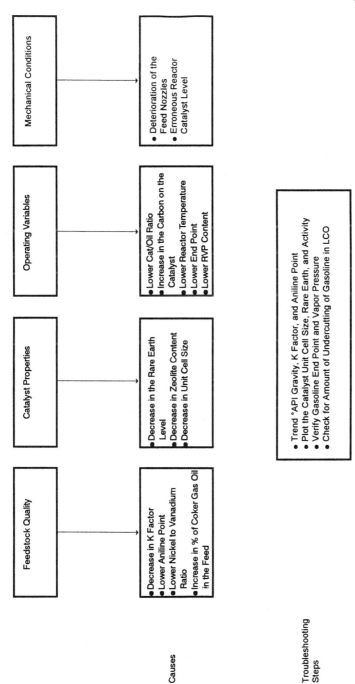

Figure 8-11C. Troubleshooting low gasoline yield.

mechanical conditions play important roles in affecting gasoline octane (Figure 8-11D).

Feedstock quality

Increases in the naphthene and aromatic fractions of the feedstock enhance the octane. These increases are indirectly indicated by:

- Increase in the refractive index.
- Decrease in the K factor and aniline point.
- Increase in the bromine number.

Catalyst properties

The fresh catalyst's chemical properties also influences the FCC gasoline octane. The gasoline octane is increased by:

- Decrease in rare earth and unit cell size.
- Decrease in sodium content.
- Increase in matrix activity.

Operating conditions

A number of operating variables can change the octane value. The factors that increase octane are:

- Increase in the reactor temperature—in general, one research octane increase per 17°F increase in the reactor temperature.
- Decrease in the catalyst-to-oil ratio.
- Increase in coke content of the regenerated catalyst.
- Increase in the regenerator temperature.
- Increase in the naphtha quench or HCO recycle.
- Decrease in the gasoline end point.
- Decrease in the gasoline vapor pressure.

Mechanical conditions

The main mechanical conditions that affect octane are the type and condition of the feed nozzles. Low-efficiency feed nozzles actually increase the gasoline octane due to promotion of thermal reactions in the mix zone. Installing "high efficiency" feed nozzles improves feed/catalyst mixing and increases the gasoline yield, but it decreases gasoline octane.

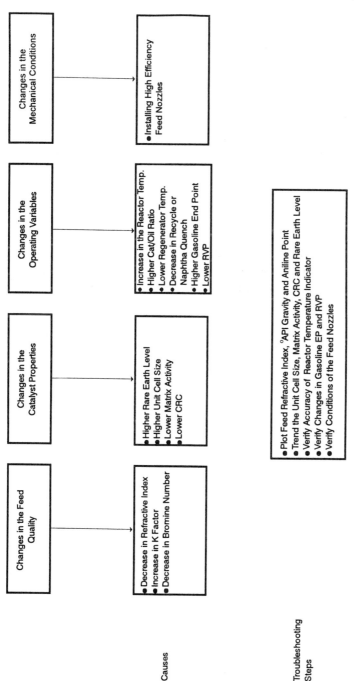

Changes in the Mechanical Conditions
- Installing High Efficiency Feed Nozzles

Changes in the Operating Variables
- Increase in the Reactor Temp.
- Higher Cat/Oil Ratio
- Lower Regenerator Temp.
- Decrease in Recycle or Naphtha Quench
- Higher Gasoline End Point
- Lower RVP

Changes in the Catalyst Properties
- Higher Rare Earth Level
- Higher Unit Cell Size
- Lower Matrix Activity
- Lower CRC

Changes in the Feed Quality
- Decrease in Refractive Index
- Increase in K Factor
- Decrease in Bromine Number

Causes

Troubleshooting Steps
- Plot Feed Refractive Index, °API Gravity and Aniline Point
- Trend the Unit Cell Size, Matrix Activity, CRC and Rare Earth Level
- Verify Accuracy of Reactor Temperature Indicator
- Verify Changes in Gasoline EP and RVP
- Verify Conditions of the Feed Nozzles

Figure 8-11D. Troubleshooting low gasoline octane.

Troubleshooting steps

- Plot the feed refractive index, °API gravity, and aniline point. Determine any shift in the amount of cracked gas oil in the feed.
- Track the unit cell size, matrix activity, and rare earth content of the catalyst.
- Determine if coke on the catalyst has changed.
- Verify accuracy of the reactor temperature.
- Check for changes in the gasoline end point and vapor pressure.
- Check the conditions of the feed nozzles and amount of atomizing steam.

SUMMARY

This chapter highlights the common problems, symptoms, and probable causes that one may encounter in troubleshooting FCC units. In addition, a systematic approach is outlined to provide solutions and corrective action.

CHAPTER 9

Emerging Trends in Fluidized Catalytic Cracking

Although the demand growth for transportation motor fuels in the North American continent is projected to be limited, the economic growth in other parts of the world will require crude oil based fuels to sustain this growth. The Far East, Latin America, and the former Soviet Union are prime examples of areas where there will be substantial demand for transportation fuels. The collapse of communism, the privatization of state-owned oil companies, and the global awareness of "environmentally clean fuels" will be some reasons for this resurgence.

In the coming years, the refining industry will be experiencing major challenges. This is especially true in the United States, in which refiners are faced with excess refining capacity, projected slow growth, and high capital and operating costs to comply with environmental, health, and safety regulations. However, the oil industry in general and the refining industry in particular are technologically sophisticated industries. They have a long history of innovations and proven track records in responding to the challenge.

It is likely that the reliable crude oil supply will not diminish any time soon. Petroleum-derived fuels will remain the primary source of transportation energy for well into the twenty-first century. Producers and refiners have been and will be environmentally competitive. The existing infrastructure of advanced product distribution systems can compete with alternative fuels readily. It is important that the future fuels be competitive, both economically and environmentally. New global market conditions will dictate closure of inefficient facilities and the investment of new technology in the larger and more efficient operations, resulting in a greater focus on the "niche market."

Optimum performance and reliability of FCC units will play an important role in the competitiveness and survival of refineries. Over the years, the FCC has proven to be a versatile process meeting the needs and demands of refiners. As one of the most efficient conversion processes in the refinery, its versatility and high degree of efficiency will continue to play a key role in meeting future reformulated fuel demands.

The objective of this chapter is to discuss:

- Evolution of reformulated fuels and its impact on FCC operations.
- Resid upgrading through FCC.
- Gaseous emissions from FCC and the various technologies that reduce these emissions.
- Emerging developments in catalyst, process, and hardware technologies.

9.1 REFORMULATED FUELS

The passage of the Clean Air Act Amendment (CAAA) on November 15, 1990, has set a process for regulating the composition and quality of gasoline and diesel fuels sold in the United States. The CAAA's intent is to improve the nation's air quality by reducing ozone and other air pollutants. Title II of the CAAA requires the manufacture and sale of "cleaner" fuels in order to reduce evaporative and combustible emission of:

1. Volatile organic compounds (VOCs).
2. Nitrogen oxides (NO_x).
3. Toxins in which benzene is the predominant concern, followed by formaldehyde, acetaldehyde, 1,3 butadiene, and poly aromatics material (pom).

9.1.1 VOCs Emission

The evaporative emissions of gasoline are mainly due to the presence of butane and the low-boiling light olefins (C_4 and C_5). Reducing gasoline vapor pressure and removing these olefins can limit the amount of evaporative emissions. Light olefins are photochemically reactive; removing them will improve ozone.

The engine operating mode controls the tailpipe emissions of hydrocarbons (HC) and carbon monoxide (CO). It is during the cold engine and warmup operation that over 80% of HC and CO emissions are generated. Fuel vaporization and fuel/air mixing are important factors in achieving a thorough combustion of the hydrocarbons.

Gasoline properties can be modified to be vaporized quickly and fully. This is often accomplished by:

- Decreasing the end point or 90% boiling point.
- Reducing the aromatic content.
- Adding oxygenates.

The high-boiling-point temperature of a fuel contributes to its partial vaporization in a cold-engine operation. Reducing the 90%-point temperature of the gasoline or lowering the 50%-point temperature will reduce HC emissions in the engine exhaust.

Aromatic levels and carbon content of the gasoline have also a significant effect on the tailpipe emissions of HC and CO. Because of their high heat of vaporization and high boiling point (see Figure 9-1), aromatics as compared to other gasoline-type hydrocarbons do not vaporize readily. Thus, there exists incentive to minimize aromatics.

Oxygenates reduce CO emissions by "enleaning" the fuel-to-air mixture for combustion. Enleanment of the fuel with oxygenates has the most impact on CO emissions. However, oxygenates, particularly ethers, are often used as a "substitute" for aromatics in achieving octane specifications. This "backing out" of aromatics further reduces CO and HC emissions.

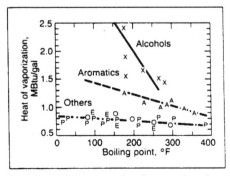

X = alcohols, A = aromatics, P = paraffins, O = olefins and E = ethers.

Figure 9-1.   Heat of vaporization versus boiling point [16].

9.1.2 NO$_x$ Emission

Similar to CO, direct exposure to NO$_x$ can cause respiratory problems. Additionally, oxidization of NO$_x$ to nitrogen dioxide in the atmosphere can contribute to acid rain. Furthermore, NO$_x$ can increase the levels of ozone and smog.

The main stationary sources of NO$_x$ are gas turbines, fired heaters, and power generation plants using coal and heavy fuel oil. The amount of NO$_x$ produced is a function of time and combustion temperature. Combustion temperature is influenced by fuel composition.

The mobile source of NO$_x$ is the combustion of fuel in an internal combustion engine. Because aromatics have the highest combustion temperature among the other hydrocarbon types (see Figure 9-2), they tend to produce higher amounts of NO$_x$ in the exhaust gases than olefins and paraffins.

One might think that because oxygenates have lower combustion temperatures, they are expected to generate less NO$_x$ than gasoline fuel without the oxygenates. On the contrary, because of the enleanment effect that raises combustion temperature, oxygenates actually increase NO$_x$ emissions. Consequently, some compromise may be needed with respect to oxygen content of fuels and its effect on HC, CO, and NO$_x$ emissions.

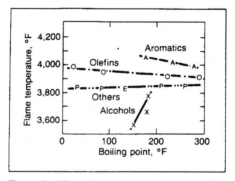

Theoretical flame temperatures assuming adiabatic and stoichiometric air.

Figure 9-2. Flame temperature of aromatics, olefins, paraffins, and alcohol [16].

9.1.3 Benzene Emission

Benzene is a known carcinogen. The U.S. Environmental Protection Agency (EPA) has identified benzene as a toxic air pollutant (TAP). Benzene is present in automotive evaporation, refueling vapors, and exhaust. Exhaust benzene is a function of aromatics and benzene content. The formula to calculate exhaust benzene emission is:

$$EXB = 1.884 + 0.949 \times BZ + 0.113 \times (A - BZ)$$

Where:

EXB = Exhaust benzene, milligram/mile
BZ = Benzene, vol%
A   = Aromatics, vol%

For example, the exhaust benzene for a gasoline having 30 vol% aromatics and 1 vol% benzene is = 6.11 milligram/mile.

9.1.4 CAAA Regulations

As of November 1, 1992, all gasoline sold in the 39 CO nonattainment areas must contain 2.7 wt% oxygen during the winter months. In addition, beginning in January 1, 1995, the regulations mandate that the gasoline sold in the nine worst ozone nonattainment metropolitan areas (Los Angeles, New York, Greater Connecticut, Baltimore, Philadelphia, Chicago, Milwaukee, Houston, and San Diego) must contain at least 2.0 wt% oxygen and not more than 1 vol% benzene and 25 vol% total aromatics. Other cities which have had mobile-source emission problems can "opt-in" voluntarily to the use of reformulated fuels.

From 1995 through 2000, there should be a 15% reduction in VOCs and other air toxins in the nonattainment areas. Effective in the year 2000, the required reduction could be 25% of the baseline unless the EPA determines that too costly or not feasible. In the latter case, the reduction could not be less than 20%.

The oxygen is added as oxygenated hydrocarbon components, specifically methyl tert-butyl ether (MTBE), tert-amyl methyl ether (TAME), ethyl tert-butyl ether (ETBE), di-isopropyl ether (DIPE), ethanol, methanol, and tertiary butyl alcohol (TBA). The properties of oxygenates as they relate to gasoline blending is shown in Table 9-1.

The key points of the regulations will also require:

- The certification of fuels, where each refiner, blender, or importer of gasoline must ensure that per-gallon emissions levels of VOCs, NO_x, CO, and toxic air pollutants do not exceed the gasoline sold in 1990.
- Effective October 1993, that highway diesel fuel must have a maximum sulfur content of 0.05 wt% (500 ppm) and a minimum cetane rating of 40.
- The use of engine additives in all gasoline to prevent accumulation of deposits in engines or fuel supply systems.
- Elimination of lead and lead additives by the end of 1995.

The standards for conventional or non-RFG gasoline are shown in Table 9-2. Oil companies can choose to comply with the requirements of conventional gasoline on either a per-gallon baseline or the company's 1990 baseline. If there is an incremental volume of fuel above the 1990 production rate, the baseline will be adjusted using the industry's baseline data (see Table 9-3). The industry's baseline gasoline is an average of properties of all the U.S. gasoline marketed in 1990.

Table 9-1
Oxygenates Properties

	MTBE	ETBE	TAME	TBA	Ethanol	Methanol
Blending Octane (R + M)/2	110	111	105	100	115	108
Blending RVP, psi	8	4	1		18	31+
Boiling point, °F	131	161	187	181	173	148
Density @ 60°F, lb/gal	6.2	6.2	6.4	6.6	6.6	6.6
Water Solubility, wt%	1.4	0.6	—			
Max. Concentration, vol%	15.0	12.7	12.4	16.1	10	9.7
Maximum Oxygen, wt%	2.7	2.0	2.0	3.7	3.7	3.7

Source: Piel [16]

The *Simple Model* RFG requires the addition of oxygenates and it limits the amount of benzene, sulfur, olefins, and T_{90}. The RVP is also lowered for six months during the summer period. Given these requirements, companies can choose to comply on a per-gallon basis (Table 9-4) or adopt the 1990 industry average basis (Table 9-5).

Table 9-2
Conventional Gasoline Standards

Properties	Specifications
Exhaust Benzene, mg/mile	Maximum 100% of baseline
Sulfur, ppmw—yearly average	Maximum 125% of baseline
Olefins, vol%—yearly average	Maximum 125% of baseline
T_{90}, °F—yearly average	Maximum 125% of baseline

Table 9-3
U.S. Industry 1990 Baseline
for Non-RFG Gasoline

Aromatics, vol%	28.6
Olefins, vol%	10.8
Benzene, vol%	1.6
Sulfur, ppmw	338
Exhaust Benzene, mg/mile	6.45
T_{90}, °F	332

Table 9-4
RFG Simple Model per Gallon Standards

RVP	
VOC Control Region 1 (South)	7.2 psi, maximum
VOC Control Region 2 (North)	8.1 psi, maximum
Oxygen content, wt%	2.0–2.7
Toxics reduction	15.0%, minimum
Benzene, vol%	1.00%, maximum
Sulfur, ppmw—yearly average	100% baseline, maximum
Olefins, vol%	100% baseline, maximum
T_{90}, °F	100% baseline, maximum

Table 9-5
RFG Simple Model Average Gasoline Standards

RVP	VOC-Control Region 1 (South)
	Standard: 7.1 psi, max.
	Per-gallon: 7.4 psi, max.
	VOC-Control Region 2 (North)
	Standard: 8.0 psi, max.
	Per-gallon: 8.3 psi, max
Oxygen Content, wt%	Standard: 2.1–2.7
	Per-gallon: 1.5–2.7
Toxics Reduction	16.5%, min.
Benzene, vol%	Standard: 0.95, maximum
	Per-gallon: 1.30, maximum
Sulfur, ppmw—yearly average	100% baseline, maximum
Olefins, vol%—yearly average	100% baseline, maximum
T_{90}, °F—yearly average	100% baseline, maximum

Starting January 1998, the EPA's *Complex Model* will go into effect. The Complex Model provides a set of equations that predict VOC, NO_x, and toxic emissions. These equations use eight gasoline properties to predict exhaust and nonexhaust emissions. These properties are RVP, oxygen, aromatics, benzene, olefins, sulfur, E200, and E300. E200 and E300 are the percent of gasoline evaporated at 200°F and 300°F, respectively.

The Complex Model contains the following:

- Seven exhaust emission equations for VOCs and NO_x, and five toxins (benzene, butadiene, formaldehyde, acetaldehyde, and polycyclic organic material (POM).
- Four nonexhaust emission equations for VOCs (diurnal, hot soak, running loss, and refueling emissions).
- Four corresponding nonexhaust emission equations for benzene.

These *nonlinear* equations can be embedded into the refinery's linear programming (LP) to achieve compliance and optimize the gasoline blend.

The key FCC gasoline components which influence RFG are:

- Sulfur
- Benzene and aromatics
- Olefins

9.1.5 Sulfur

Sulfur in gasoline is a contributor to the SO_x air quality problem. Sulfur also deactivates the catalyst of the automotive's catalytic converter and reduces its efficiency. The sulfur compounds in FCC gasoline consist of C_1–C_4 mercaptan and various thiophenes.

The California Air Resources Board (CARB) has set an average sulfur specification of 40 ppm for 1996, with a maximum of 80 ppm. The CAAA's Complex Model also addresses sulfur issues in its set of equations.

FCC gasoline is by far the largest sulfur contributor (up to 90%) in the gasoline pool. Therefore, controlling gasoline sulfur involves reducing sulfur content of the FCC gasoline. There are several available options to reduce the FCC gasoline's sulfur, including:

- FCC feed hydrotreating.
- Gasoline end point reduction.
- FCC gasoline hydrotreating.
- Use of catalyst additives.
- Biocatalytic desulfurization of the FCC gasoline.

Approximately 10% of the sulfur in nonhydrotreated FCC feed ends up in the FCC gasoline as compared to about 5% for hydrotreated feed. For example, if the sulfur content of a nonhydrotreated feed is 1.0%, the sulfur of the FCC gasoline will be 1,000 ppm. Assuming an 80% desulfurization, feed to the FCC unit will contain 0.2% sulfur, resulting in FCC gasoline containing 100 ppm sulfur.

Hydrotreating or moderate pressure hydrocracking of the FCC feed provides many benefits. Besides reducing sulfur in the FCC gasoline, FCC feed hydrotreating reduces NO_x and SO_x emissions from the regenerator flue gas, increases gasoline yield, and reduces catalyst consumption rate. However, a number of refiners cannot justify the high capital cost of FCC feed hydrotreating.

Reducing the gasoline end point can significantly decrease the FCC gasoline sulfur (Figure 9-3); about 50% of sulfur is contained in the last 5 vol%. For those refiners processing a high sulfur crude mix, this end point reduction may not be sufficient to meet sulfur specifications. The disposition of this high-sulfur, high-aromatic gasoline can be a problem. One option is to combine this heavy fraction with the LCO stream and desulfurize it in the diesel hydrotreater. After

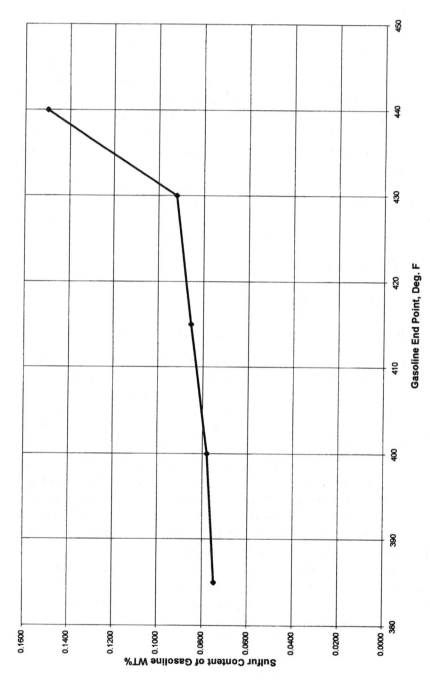

Figure 9-3. Gasoline sulfur versus its end point.

hydrotreating, the heavy gasoline can be separated and sent to the gasoline pool.

Selective hydrodesulfurization (HDS) of the FCC gasoline can prove to be a positive choice for meeting the required sulfur levels. The major drawback of FCC gasoline desulfurization is octane loss as a result of olefins saturation. An HDS process can be designed to treat either the full-range FCC gasoline cuts or the heavier cuts to minimize olefin saturation and octane loss. The choice of a proper catalyst and operating conditions is important in maximizing sulfur reduction and minimizing octane loss.

Catalyst additives can be an option in reducing FCC gasoline sulfur. They can reduce the gasoline sulfur by about 15% and they work by converting mercaptan, thiophene, etc., to H_2S. A secondary benefit of the additives is an approximate 10% reduction in the LCO sulfur.

Biocatalytic desulfurization (BDS) technology employs the concept of enzymatic removal of sulfur compounds without changing the fundamental structure of hydrocarbons. Energy Biosytem of Houston is developing the process in conjunction with Petrolite Corporation.

9.1.6 Aromatics and Benzene

The tailpipe emissions of HC and CO is affected by the levels of heavy aromatics in gasoline. Like sulfur, the heavy aromatics are in the back end of the boiling range (Figure 9-4). As with sulfur, reduction of end point directly controls the concentration of heavy aromatics in finished gasoline.

The benzene content of FCC gasoline is typically in the range of 0.6 to 1.3 vol%. The CAAA's simple model requires RFG to have a maximum of 1 vol% benzene. In California, the basic requirement is also 1 vol%, however, if refiners are to comply with averaging provisions, then the maximum is 0.8 vol%. Operationally, the benzene content of FCC gasoline can be reduced by reducing catalyst-oil contact time and catalyst-to-oil ratio. Lower reactor temperature, lower rates of hydrogen transfer, and using an "octane catalyst" will also reduce benzene levels.

Most of the benzene in the gasoline pool comes from the reformer unit (reformate). To reduce the reformate's benzene, one must modify the feedstock quality and/or operating conditions. Benzene's precursors in the reformer feed (C_5 and C_6) can be prefractionated and sent to

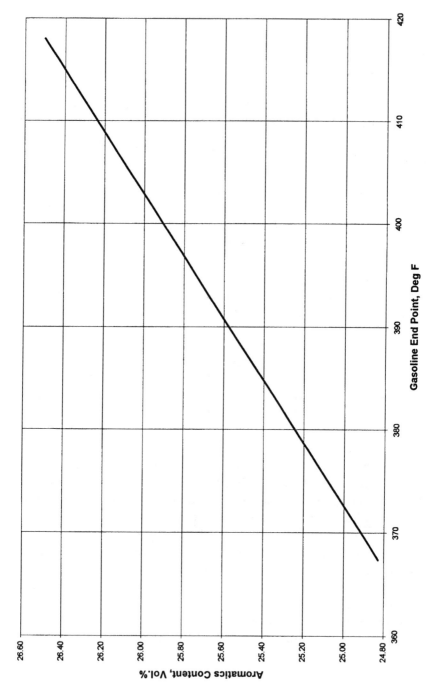

Figure 9-4. Aromatic content of gasoline versus its end point.

an isomerization unit. The reformer operating pressure can be reduced to reduce benzene and aromatics. Another option is postfractionation of the reformate stream. This requires the installation of a reformate splitter. The light aromatics (benzene, toluene, and mix xylene) in the "light" reformate can be extracted for petrochemical grade operations. Toluene can be converted to benzene through the hydrodealkylation process. The benzene can be saturated to cyclohexane and eventually isomerized. A combination of the benzene saturation system and paraffin isomerization will enable the refiner to control benzene while improving the gasoline octane pool.

9.1.7 Olefins

In 1990, U.S. gasoline contained about 10 vol% olefins. The majority of these olefins emanate from FCC gasoline. FCC gasoline has in the range of 25–35 vol% olefins. For non-RFG gasoline, as with sulfur, the regulation allows the maximum olefin content to be 125% of the 1990 baseline values.

Light olefins are very reactive to forming ozone. However, the future trend of most FCC operations is projected to produce more and more olefin feed. This is because olefins, particularly the C_4 and C_5 olefins, can either be "alkylated" and/or "etherified." The propylene can also be alkylated or used for petrochemical feedstock.

Commercial alkylation is the reaction of isobutane with C_3 through C_5 olefins in the presence of either sulfuric acid or hydrofluoric acid (see Example 9-1). In the presence of an acidic catalyst, ethers are produced by reacting an alcohol with an olefin having a double bond on a tertiary carbon atom (see Example 9-2).

Example 9-1
Isobutane Alkylation of Propylene and Butylene

Propylene Alkylation

$$CH_3 - CH = CH_2 + CH_3 - \underset{\underset{CH_3}{|}}{CH} - CH_2 \rightarrow CH_3 - \underset{\underset{CH_3}{|}}{CH} - \underset{\underset{CH_3}{|}}{CH} - CH_2 - CH_3$$

PROPYLENE + ISOBUTANE → DIMETHYLPENTANE

Typical Yield:

1.0 Volume of Propylene + 1.3 Volume of Isobutane → 1.80 Volume of Alkylate.

Butylene Alkylation

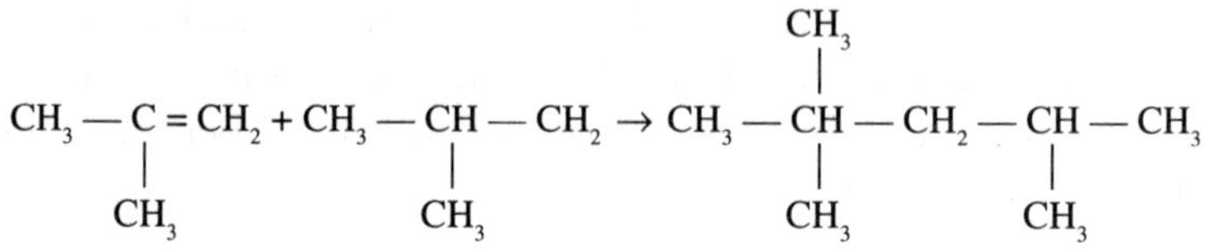

$$CH_3 - C = CH_2 + CH_3 - CH - CH_2 \rightarrow CH_3 - \overset{\overset{\displaystyle CH_3}{|}}{CH} - CH_2 - CH - CH_3$$

$$\underset{CH_3}{|} \qquad \underset{CH_3}{|} \qquad \underset{CH_3}{|} \qquad \underset{CH_3}{|}$$

PROPYLENE + ISOBUTANE → 2,2,4 TRIMETHYLPENTANE

Typical Yield:

1.0 Volume of Butylene plus 1.2 Volume of Isobutane Yield 1.70 Volume of Alkylate.

Example 9-2
Etherification of Methanol

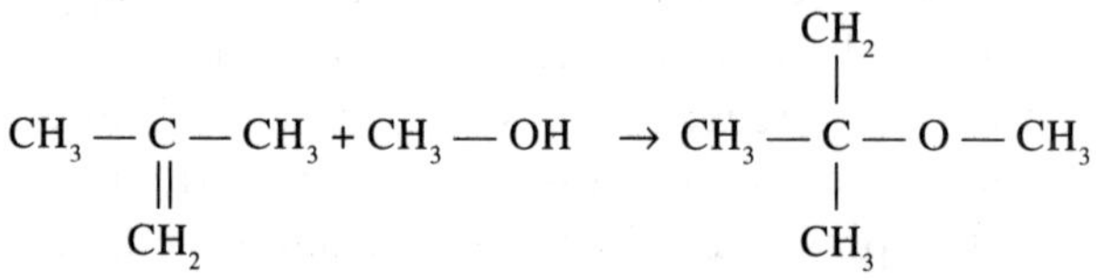

$$CH_3 - \overset{\overset{\displaystyle |}{C}}{\underset{\overset{\displaystyle ||}{CH_2}}{}} - CH_3 + CH_3 - OH \rightarrow CH_3 - \overset{\overset{\displaystyle CH_2}{|}}{\underset{\overset{\displaystyle |}{CH_3}}{C}} - O - CH_3$$

ISOBUTYLENE + METHANOL → METHYL TERTIARY BUTYL ETHER (MTBE)

Typical Yields:

1.0 Volume of Isobutane plus 0.43 Volume of Methanol Yields 1.27 Volume of MTBE.

Both alkylate and ether have excellent properties as gasoline blending components. A very low RVP, a high road octane, no aromatics, and virtually zero sulfur are some of their superior properties. The emphasis on alkylation and etherification will continue in both the U.S. and the rest of the world.

A conventional FCC unit can be an "olefin machine" with proper changes in catalyst, operating conditions, and hardware. FCC catalysts having a low unit cell size and a high silica/alumina ratio favor olefins. Additionally, the addition of ZSM-5, with its lower acid site density and very high-framework silica-alumina ratio, convert C_7+ gasoline into olefins. A high reactor temperature coupled with elimination of the post-riser residence time will also produce more olefins. Mechanical modifications of the FCC riser for "millisecond" cracking has shown potential for maximizing olefin yield.

9.1.8 Challenges Facing RFG

RFG is a cost-effective fuel which improves air quality and is a mechanism through which the refining industry can be competitive. In the years to come, numerous issues regarding RFG will be facing refiners. Most are regulatory, political, and bureaucratic issues. Following are some of these issues regarding RFG:

- Public perception of RFG regarding health effects of ethers, and price increase and engine performance complaints.
- EPA's ethanol mandate and the subsequent stay of that mandate by the federal court.
- Complexity of testing, distribution, storage, handling, and blending facilities.
- Record keeping and the development of a uniform certification program.
- Interchangeability of MTBE to ETBE.
- Interpreting the baseline.
- The future of opt-in areas: the continual decline in air quality where RFG is not sold.
- antidumping, credits, and trading.
- The program length of oxygenated fuels for CO nonattainment areas.
- The definition of "domestic supply."

9.2 RESIDUAL FLUIDIZED CATALYTIC CRACKING

Deterioration in worldwide crude oil supply (Table 9-6), continual decline in the demand of heavy fuel oil, and the recent mechanical and catalyst advances have provided potential economic incentives to

Table 9-6
U.S. Crude Characteristics

Year	°API Gravity	Wt% Sulfur
1983	32.92	0.88
1984	32.96	0.94
1985	32.46	0.91
1986	32.33	0.96
1987	32.22	0.99
1988	31.93	1.04
1989	32.14	1.06
1990	31.86	1.10
1991	31.64	1.13
1992	31.32	1.16
1993	31.30	1.15

Source: Swain [24]

upgrade the atmospheric and/or vacuum bottoms in the *residual fluid-ized catalytic cracking* (RFCC) unit. Although the residual upgrading in the United States is mostly a delayed coker based bottoms upgrading, most of the new FCC units are either residue crackers or have in-place provisions to process residue at a later date. This is more pronounced in the new units built in the Far East, Europe, and Australia. The main reason is that the residue from their crude oils tends to be more paraffinic and contain less metals than North Sea or Middle Eastern crude oils, which makes them less suitable for RFCC.

The main difference that distinguishes an RFCC from a conventional vacuum gas oil FCC is the quality of the feedstock. The residue feed has a high coking tendency and an elevated concentration of contaminants.

9.2.1 Coking Tendency

Residue feedstocks have a higher coking tendency, which is indicated by higher levels of Conradson carbon and higher boiling point. The common definition of residue is the fractions of the feed which boil above 1050°F and Conradson carbon levels of greater than 0.5 wt%. The residual portion of the feed contains hydrogen-deficient asphaltenes and polynuclear compounds. Some of these compounds will lay down on active catalyst sites as coke, reducing catalyst activity and selectivity.

9.2.2 Feed Contaminants

The residual portion of feedstocks also contains a large concentration of contaminants. The major contaminants, mostly organic in nature, are nickel, vanadium, nitrogen, and sulfur. Nickel, vanadium, and sodium are deposited quantitatively on the catalyst. This deposition poisons the catalyst, accelerating its production of coke and light gases.

The nickel in the feed is deposited on the surface of the catalyst, promoting undesirable dehydrogenation and condensation reactions. These nonselective reactions increase gas and coke production at the expense of gasoline and other valuable liquid products. The deleterious effects of nickel poisoning can be reduced by the use of antimony passivation.

Vanadium and sodium neutralize catalyst acid sites and can cause collapse of the zeolite structure. Figure 9-5 shows the deactivation of the catalyst activity as a function of vanadium concentration. Destruction of the zeolite by vanadium takes place in the regenerator where the combination of oxygen, steam, and high temperature forms vanadic acid according to the following equations:

$$4V + 5\ O_2 \rightarrow 2\ V_2O_5$$

$$V_2O_5 + 3\ H_2O \rightarrow 2\ VO\ (OH)_3$$

The produced vanadic acid, $VO(OH)_3$, is mobile. Sodium tends to accelerate the migration of vanadium into the zeolite. This acid attacks the catalyst, causing collapse of the zeolite pore structure.

The presence of increased basic nitrogen compounds, such as pyridines and quinoline in the FCC feedstock, also attacks catalyst acid sites. The result is a temporary loss of catalyst activity and a subsequent increase in coke and gas yields. Additionally, in the regenerator, the adsorbed nitrogen is converted to nitrogen oxide (NO_x).

Although an increase in the sulfur content of the residue feedstock will have a minimal effect on unit yields, the sulfur content of the RFCC products and the flue gas is greater, requiring additional treating facilities.

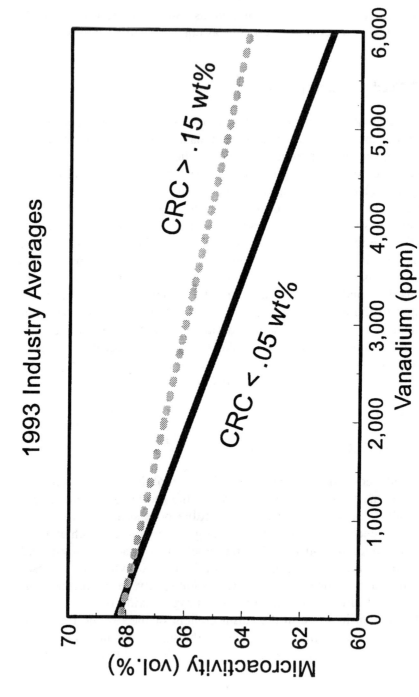

Figure 9-5. Vanadium deactivation varies with regenerator severity [25].

9.2.3 Operational Impacts of RFCC

Operationally, processing residue feedstocks has the following effects:

- The higher delta coke and the coke yield, which are associated with residue feedstocks, will result in elevated regenerator temperature and higher combustion air requirements.
- Exposure of the catalyst to a variety of feed contaminants and the higher regenerator temperature will reduce both catalyst selectivity and activity.
- The greater levels of nitrogen and sulfur in the residue feed increase the emissions of the NO_x and SO_x from the regenerator.

9.2.4 Minimizing Detrimental Effects of Processing Residual Feeds

The proper choice of a feed injection system, regenerator, and catalyst are some of the key aspects of successful RFCC operation.

An efficient feed injection system produces rapid vaporization of the feed and minimizes the amount of nonvaporized hydrocarbons that block the active sites. An effective feed nozzle system must instantaneously vaporize and crack the asphaltenes and polynuclear aromatics to lower boiling entities.

The regenerator design, either single-stage or two-stage, should provide a uniform catalyst regeneration, increase flexibility for processing a variety of feedstocks, and minimize thermal and hydrothermal deactivation of the catalyst.

The catalyst design should also be optimized to achieve the following objectives:

- Low coke and gas production.
- Efficient bottoms cracking.
- Improved metals resistance.
- Improved thermal and hydrothermal stability.
- An active matrix and a low hydrogen transfer activity to convert the bottoms and minimize delta coke.

9.3 REDUCING FCC EMISSIONS

The gaseous emissions from the FCC unit, which at the present time are either locally or nationally regulated, are CO, NO_x, particulates, and SO_x. Table 9-7 shows the current allowable limits of the EPA's New Source Performance Standards (NSPS) for the emissions of these airborne pollutants. NSPS levels can be triggered by one of the following conditions:

- Construction of a new unit.
- Revamp of the regenerator, provided the modification costs are more than 50% of a comparable regenerator.
- Any capital modification of the unit which increases its emission rates.

Although, at present, there is not a national requirement to limit NO_x emissions from the FCC flue, several state and regional agencies have imposed limits on their release. The quantity of these emissions is directly proportional to the quality of FCC stocks, operating conditions, catalyst type, and the mechanical conditions of the unit. Processing feeds that contain a high concentration of residue, sulfur, nitrogen, and metals will release a greater amount of SO_x, NO_x and particulates. The following discusses various available technologies to reduce flue gas emissions.

Table 9-7
EPA's New Source Performance Standards (NSPS) for Gaseous Emissions from the FCC Regenerators

Source	Allowable Limits
Carbon Monoxide (CO)*	Less than 500 ppmv in the flue gas
Nitrogen Oxides (NO_x)	None
Particulates**	A maximum of 1.0 pound of solids in the flue gas per 1,000 pounds of coke burned
Sulfur Oxides (SO_2 + SO_3)*	Exempt if the feed sulfur is less than 0.30 wt%.
	If there is no add-on control such as a wet gas scrubber, 9.8 kilograms of (SO_2 + SO_3) per 1,000 kilograms of coke burned. This is approximately equal to 500 ppmv.
	Add-on device: reduce (SO_2 + SO_3) by at least 90% or not more than 50 ppmv, whichever is less stringent

*Effective January 1984
**Effective June 1973

9.3.1 Particulates

Electrostatic precipitators (ESP) and wet gas scrubbers (WGS) are widely used to remove particulates from the FCC flue gas. Both can recover over 80% of filtrable solids. An ESP (Figure 9-6) is typically installed downstream of the flue gas heat recovery (prior to atmospheric discharge) to minimize particulate concentration. If a combination of low particulate and low SO_x requirements are to be met, a wet gas scrubber such as the one licensed by Exxon (Figure 9-7) should be considered. If SO_x removal is not a prime objective, an ESP will be less expensive from the standpoints of both initial capital and operating costs. In some cases, a bag house system can be used instead of an ESP.

9.3.2 SO_x

There are three widely used methods to reduce SO_x emissions from the FCC flue gas. These are:

- FCC feed pretreatment.
- Catalyst additives.
- Flue gas desulfurization.

Feed hydrotreating or hydrocracking reduces sulfur content of FCC products, including SO_x emissions. As discussed earlier in this chapter, there are many benefits associated with FCC feed hydrotreating. It is important to note that most of the sulfur in a hydrotreated feed is in organic compounds, and therefore it will mostly be concentrated in the decanted oil and coke. Consequently, for a given sulfur in the feed, there will be more SO_x produced with the hydrotreated feed.

For refiners having low to moderate levels of SO_x in their FCC flue gas (less than 1,000 ppm), SO_x additives are the most economical methods of reducing SO_x emissions. These additives are injected separately into the regenerator. SO_x additives work by capturing SO_3 in the regenerator and releasing sulfur as H_2S in the reactor. Having a reliable SO_2 on-line analyzer will ensure that a sufficient quantity of additive is injected. Operating conditions of the regenerator, especially partial vs. full combustion and excess oxygen level, will greatly influence the additive's effectiveness.

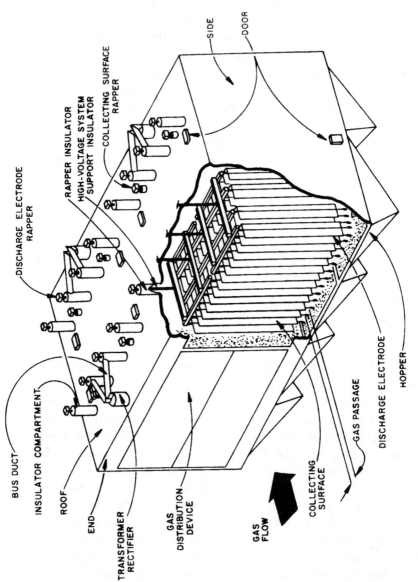

Figure 9-6. Typical electrostatic precipitator (ESP).

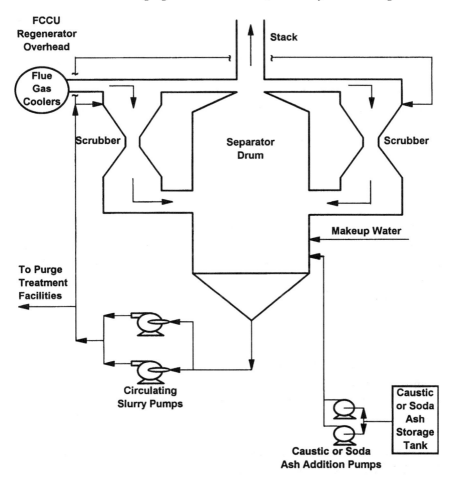

Figure 9-7. Schematic of "Exxon scrubbing system" [26].

When processing high-sulfur feeds (greater than 1.0 wt%) or if the required SO_x reduction levels are greater than 80%, other capital-intensive desulfurization technologies must be considered. There are several flue gas desulfurization technologies being offered, including Exxon Wet Gas Scrubber, Haldor Tosoe's WSA, and United Engineers' Mgo. The Exxon wet gas scrubbing is one the most widely used processes to remove SO_x. As stated earlier, the WGS process removes both SO_x and particulates from the FCC flue gas. The Exxon-licensed WGS uses a high-energy venturi (HEV) system to provide liquid/gas contacting.

9.3.3 CO

The CO levels released from the regenerator flue gas operating either in complete or partial combustion are normally less than 100 ppm. For units operating in partial combustion, the flue gas is often sent to a CO boiler(s) prior to its discharge to the atmosphere. The concentration of CO in the flue gas exiting the regenerator depends largely on the operating conditions of the regenerator and the CO promoter and the efficiency of the air/spent catalyst distribution system.

9.3.4 NO_x

The NO_x levels in the FCC flue gases ranges from 50–500 ppm. Nitrogen content of the feed, excess oxygen, regenerator residence time, dense phase temperature, and CO promoter all influence the concentration of NO_x in the flue gas. In the regenerator, the nitrogen oxide reacts with CO according to the following equation:

$$2NO + CO \rightarrow N_2 + CO_2 + 1/2 \; O_2$$

This indicates that a large portion of the NO_x in the flue gas originates from the oxidation of inert nitrogen (in the presence of CO promoter and excess oxygen) and not from feed nitrogen. Ammonia can be used (over a catalyst) to neutralize NO_x to produce nitrogen and water.

9.4 EMERGING DEVELOPMENTS IN CATALYSTS, PROCESSES AND HARDWARE

The FCC process has a long history of innovation; it has been and will continue to play a key role in the overall success of the refining industry. The continuing developments will primarily be in the areas of catalyst, process, and hardware technologies.

9.4.1 Catalyst

Since the mid 1960s, there has been a steady improvement in formulation of the FCC catalysts. The focus and emphasis of the research is in the following areas:

- Improvement in zeolite quality.
- Increase in the quantity and choice of active matrix.

- Customization of the catalyst to the unit objectives and constraints.
- A widespread use of ZSM-5 or similar zeolite.

There has also been an ongoing trend to formulate a higher-quality zeolite. Higher quality has been reflected in:

- Greater silica-to-alumina (SAR) of zeolite—a greater SAR results in a zeolite which is more stable, yields more olefins, improves octane, and increases product selectivity.
- Improved crystallinity by producing zeolite crystals which are more uniform. FCC catalyst manufacturers have greater control over the zeolite acid sites distribution. In addition, there is an upward trend in the quantity of zeolite in the catalyst.

The selectivity and activity of the catalyst matrix will continue to improve. The emphasis in bottoms cracking and a steady reduction in the reaction residence time has demanded an increase in the quantity of active matrix.

Future catalyst formulation will be customized to meet the individual refiner's needs moreso than in the past. Catalyst manufacturers will be tailoring catalysts to meet each refiner's requirements.

The demand for ZSM-5 additives will increase because of their inherent ability to crack low-octane, straight-chain olefins to C_3 and C_4 olefins and also to isomerize low-octane linear olefins to higher octane branched olefins. Additionally, once ZSM-5's patent has expired, its use should increase.

9.4.2 Operating Conditions

FCC will still play a dominant role in producing cleaner-burning fuels. The inherent flexibility of the process will allow refiners to meet the fuel reformulation requirements. With the anticipated growing demand for alkylate and ethers, the FCC operating parameters will be adjusted to maximize production of propylene, isobutylene, and iso-amylene. The projected trend in operating conditions will therefore be in a higher reactor temperature, a higher catalyst-to-oil ratio, a higher reaction mix temperature, and shorter catalyst contact.

9.4.3 Technology Development

Since 1942, when the first FCC unit came onstream, new technologies have continuously evolved to maximize the unit's performance and to meet the ever-changing product requirements and feedstock qualities. The pace of future technology development will remain dynamic. Examples of the new and ongoing technologies aimed at enhancing the unit's operational and mechanical performance as well as complying with environmental regulations are:

- Reducing sulfur and aromatics from gasoline and distillate.
- Minimizing disposal of the equilibrium catalyst.
- Minimizing catalyst back-mixing in the riser to minimize production of undesirable products. Redesign of the conventional riser for a down-flow of catalyst and vapors could virtually eliminate this back-mixing.
- Achieving an ultrashort catalyst-hydrocarbon contact time designed to maximize olefins and gasoline yields.
- Eliminating long dilute-phase residence time downstream of the riser to prevent recracking of hydrocarbon vapors in the reactor housing.
- Improving feed and catalyst injection systems.
- Improving spent catalyst distribution.
- Improving mechanical reliability of the FCC reactor-regenerator components.
- Increasing use of feed segregation to maximize production of light olefins.
- Increasing use of riser quench to maximize reaction mix temperature and to promote maximum vaporization of the feedstock.
- Increasing use of catalyst additives to reduce gaseous emissions and to maximize light olefins.

These are just some of the many challenges facing FCC operations today.

SUMMARY

The United States refining industry is undergoing a restructuring phase. Refiners will continue to be under pressure and only the most efficient and profitable operations are going to survive. The survivors are those who have some niche in the market place, have the versatility to handle low-cost crude, meet product demand, and conform with environmental regulations.

FCC is one of the cheapest conversion refining processes. Its inherent flexibility can assist a refiner in meeting changing product requirements, especially considering the steady decline in feedstock quality.

The U.S. federal RFG program has imposed new challenges for the FCC, particularly regarding the sulfur, aromatics, and olefin content of gasoline. Various commercially-proven technologies, along with evolving technologies, will be available to comply with these new rules.

The use of RFCC will continue to grow, particularly in regions of the world where the residue feedstock contains low levels of contaminants. Attention to regenerator and feed injection designs is important in ensuring a successful operation.

Gaseous emissions (CO, NO_x, SO_x, particulates) from FCC have been regulated at local and national levels. The quantity of these emissions is directly related to the quality of the FCC stocks, operating conditions, catalyst type, and mechanical conditions of the unit. Processing feeds that contain a high concentration of residue, sulfur, nitrogen, and metals will release a greater amount of SO_x, NO_x, and particulates.

In conclusion, FCC has had a long history of innovations. New technological developments will continue to emerge, optimizing its performance. Its versatility and high degree of efficiency will continue to play a key role in meeting future market demands.

REFERENCES

1. Mauleon, J. L. and Letzch, W. S., "The Influence of Catalyst on the Resid FCCU Heat Balance," presented at Katalistiks' 5th Annual FCC Symposium, Vienna, Austria, May 23–24, 1984.
2. Davis, K., and Ritter, R. E., "FCC Catalyst Design Considerations for Resid Processing—Part 2," Grace Davison *Catalagram,* No. 78, 1988.
3. Hammershaimb, H. U. and Lomas, D. A., "Application of FCC Technology to Today's Refineries," presented at Katalistiks' 6th Annual FCC Symposium, Munich, Germany, May 22–23, 1985.
4. Kool, J. M., "Commercial Experience with Resid Cracking in Conventional FCC Units," presented at the 1984 Akzo Chemicals Symposium.
5. Hood, R. and Bonilla, J., "Residue Upgrading by Solvent Deasphalting and FCC," presented at the Stone & Webster 5th Annual Meeting, Dallas, Texas, October 12, 1993.

6. Dean, R. R., Hibble, P. W., and Brown, G. W., "Crude Oil Upgrading Utilizing Residual Oil Fluid Catalytic Cracking," presented at Katalistiks' 8th Annual FCC Symposium, Budapest, Hungary, June 1–4, 1987.

7. Johnson, T. E., "Resid FCC Regenerator Design," presented at the M.W. Kellogg Co. Refining Technology Seminar, Houston, Texas, February 9–10, 1995.

8. Letzsch, W., Mauleon, J. L., Jones, G., and Dean, R., "Advanced Residual Fluid Catalytic Cracking," presented at Katalistiks' 4th Annual FCC Symposium, Amsterdam, The Netherlands, May 18–19, 1983.

9. Elvin, F. J. and Krikorian, K. V., "The Key to Residue Cracking," presented at Katalistiks' 4th Annual FCC Symposium, Amsterdam, The Netherlands, May 18–19, 1983.

10. Peeples, J. E., "The Clean Air Act, a Brave New World for Fuel Reformulation," *Fuel Reformulation,* Vol. 3, No. 6, November/December 1993.

11. Dharia, D., Brahn, M., and Letzsch, W., "Technologies for Reducing FCC Emissions," presented at Stone & Webster 5th Annual Refining Seminar, Dallas, Texas, October 12, 1993.

12. Yergin, D. and Lindemer, K., "Refining Industry's Future," *Fuel Reformulation,* Vol. 3, No. 4, July/August 1993.

13. Perino, J. O., "Blending Control Upgrade Projects," *Fuel Reformulation,* Vol. 3, No. 4, July/August 1993.

14. Clarke, R. H. and Ritz, G. P., "Method for the Analysis of Complex Mix of Oxygenates in Transportation Fuels," *Fuel Reformulation,* Vol. 3, No. 4, July/August 1993.

15. Unzelman, G. H., "NO_x," *Fuel Reformulation,* Vol. 1, No. 6, November/December 1991.

16. Piel, W. J. and Thomas, R. X., "Oxygenates for Reformulated Gasoline," *Hydrocarbon Processing,* July 1990, pp. 68–73.

17. Hirshfeld, D. S. and Kolb, J., "Minimize the Cost of Producing Reformulated Gasoline," *Fuel Reformulation,* Vol. 4, No. 2, March/April 1994.

18. Unzelman, G. H., "A Sticky Point for Refiners," *Fuel Reformulation,* Vol. 2, No. 4, July/August 1992.

19. Nocca, J. L., Forestiere, A., and Cosyns, J., "Diversify Process Strategies for Reformulated Gasoline," *Fuel Reformulation,* Vol. 4, No. 5, September/October 1994.

20. Desai, P. H., Lee, S. L., Jonker, R. J., De Boer, M., Verieling, J., and Sarli, M. S., "Reduce Sulfur in FCC Gasoline," *Fuel Reformulation,* Vol. 4, No. 6, November/December 1994.

21. Sarathy, P. R., "Profit from Refinery Olefins," *Fuel Reformulation,* Vol. 3, No. 5, September/October 1993.

22. Hostetler, R. and Cain, M., BP Oil, private communication, 1995.

23. Reid, T. A., Akzo Nobel, private communication, 1995.

24. Swain, E. J., "U.S. Crude Slate Continues to Get Heavier, Higher in Sulfur," *Oil & Gas Journal,* January 9, 1995, pp. 37–42.

25. Dougan, T. J., Alkemade, V., Lakhampel, B., and Brock, L. T., "Advances in FCC Vanadium Tolerance," NPRA Annual Meeting, San Antonio, Texas, March 20, 1994, reprinted in Grace Davison *Catalagram.*

26. Cunic, J. D., Diener, R., and Ellis, E. G., Exxon Research and Engineering, "Scrubbing—Best Demonstrated Technology for FCC Emission Control," presented at NPRA Annual Meeting, San Antonio, Texas, 1990.

APPENDIX 1

Temperature Variation of Liquid Viscosity

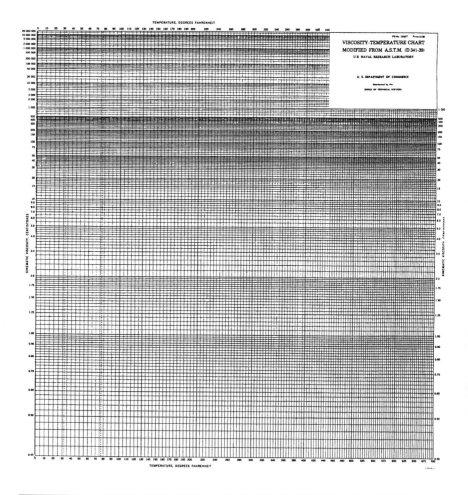

Source: U.S. Department of Commerce, adapted from ASTM D-342-39.

Correction to Volumetric Average Boiling Point

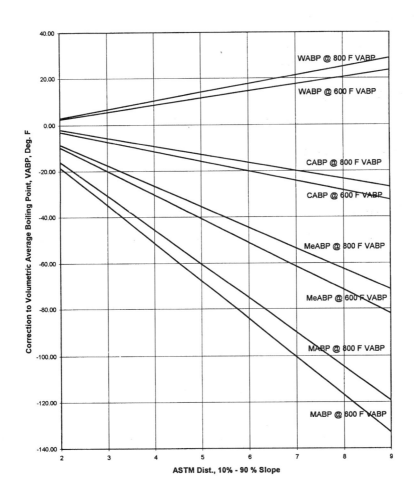

TOTAL Correlations

Aromatic Carbon Content:

$$CA = -814.136 + 635.192 \times RI(20) - 129.266 \times SG + 0.1013 \times MW - 0.340 \times S - 6.872 \times \ln(v)$$

Hydrogen Content:

$$H2 = 52.825 - 14.26 \times RI(20) - 21.329 \times SG - 0.0024 \times MW - 0.052 \times S + 0.757 \times \ln(v)$$

Molecular Weight:

$$MW = 7.8312 \times 10^{-3} \times SG^{-0.0976} \times AP_{°C}^{0.1238}$$

Refractive Index @ 20°C:

$$RI(20) = 1 + 0.8447 \times SG^{1.2056} \times (VABP_{°C} + 273.16)^{-0.0557} \times MW^{-0.0044}$$

Refractive Index @ 60°C:

$$RI(60) = 1 + 0.8156 \times SG^{1.2392} \times (VABP_{°C} + 273.16)^{-0.0576} \times MW^{-0.0007}$$

Source: Dhulesia, H., "New Correlations Predict FCC Feed Characterization Parameters," *Oil & Gas Journal*, Jan. 13, 1986, pp. 51–54.

n-d-M Correlations

$$\upsilon = 2.5 \times (RI_{20°C} - 1.4750) - (d_{20°C} - 0.8510)$$

$$\varpi = (d_{20°C} - 0.8510) - 1.11 \times (RI_{20°C} - 1.4750)$$

If υ is positive: $\%C_A = 430 \times \upsilon + \dfrac{3660}{M}$

If υ is negative: $\%C_A = 670 \times \upsilon + \dfrac{3660}{M}$

If ϖ is positive: $\%C_R = 820 \times \varpi - 3 \times S + 10,000/M$

If ϖ is negative: $\%C_R = 1440 \times \varpi - 3S + \dfrac{10,600}{M}$

$$\%C_N = \%C_R - \%C_A$$

$$\%C_P = 100 - \%C_R$$

Average Number of Aromatic Rings per Molecule (R_A):

$R_A = 0.44 + 0.055 \times M \times \upsilon$ If υ is positive

$R_A = 0.44 + 0.080 \times M \times \upsilon$ If υ is negative

Average Total Number of Rings per Molecule (R_T):

$R_T = 1.33 + 0.146 \times M \times (\varpi - 0.005 \times S)$ If ϖ is positive

$R_N = R_T - R_A$

$R_T = 1.33 + 0.180 \times M \times (\varpi - 0.005 \times S)$ If ϖ is negative

Average Number of Napthene Rings per Molecule (R_N):

$R_N = R_T - R_A$

Estimation of Molecular Weight of Petroleum Oils from Viscosity Measurements

Tabulation of H Function

	H									
	0	1	2	3	4	5	6	7	8	9
40	334	336	339	341	343	345	347	349	352	354
50	355	357	359	361	363	364	366	368	369	371
60	372	374	375	377	378	380	381	382	384	385
70	386	387	388	390	391	392	393	394	395	397
80	398	399	400	401	402	403	404	405	406	407
90	408	409	410	410	411	412	413	414	415	415
100	416	417	418	419	420	420	421	422	423	423
110	424	425	425	426	427	428	428	429	430	430
120	431	432	432	433	433	434	435	435	436	437
130	437	438	438	439	439	440	441	441	442	442
140	443	443	444	444	445	446	446	447	447	448
150	448	449	449	450	450	450	451	451	452	452
160	453	453	454	454	455	455	456	456	456	457
170	457	458	458	459	459	460	460	460	461	461
180	461	462	462	463	463	463	464	464	465	465
190	465	466	466	466	467	467	468	468	468	469

Viscosity-Molecular Weight Chart

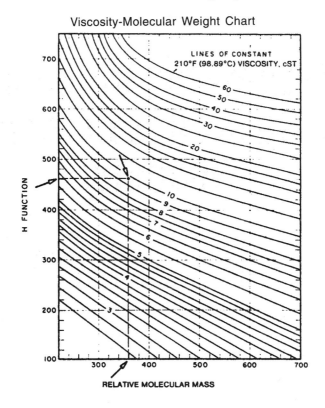

RELATIVE MOLECULAR MASS

Kinematic Viscosity to Saybolt Universal Viscosity

Kinematic Viscosity, cSt	Equivalent Saybolt Universal Viscosity, Sus	
	At 100°F	At 210°F
1.81	32.0	32.2
2.71	35.0	35.2
4.26	40.0	40.3
7.37	50.0	50.3
10.33	60.0	60.4
13.08	70.0	70.5
15.66	80.0	80.5
18.12	90.0	90.6
20.54	100.0	100.7
43.0	200.0	202.0
64.6	300.0	302.0
86.2	400.0	402.0
108.0	500.0	504.0
129.5	600.0	604.0
139.8	648.0	652.0
151.0	700.0	
172.6	800.0	
194.2	900.0	
215.8	1000.0	

Extracted from ASTM Method D-2161-87. Copyright ASTM. Used with permission.

API Correlations

$$X_p = a + b \times (R_i) + c \times (VG)$$

$$X_n = d + e \times (R_i) + f \times (VG)$$

$$X_n = g + h \times (R_i) + i \times (VG)$$

Where constants vary with molecular weight range given below:

Constants	Heavy Fractions $200 < MW < 600$
a	+2.5737
b	+1.0133
c	−3.573
d	+2.464
e	−3.6701
f	+1.96312
g	−4.0377
h	+2.6568
i	+1.60988

R_i = Refractivity Intercept
VGC = Viscosity Gravity Constant

$$R_i = R_{i(20)} - \frac{d}{2}$$

Where:

$R_{i(20)}$ = Refractive Index @ 20°C
d = Density @ 20°C

Source: Riazi, M. R., and Daubert, T. E., "Prediction of the Composition of Petroleum Fractions," *Ind. Eng. Chem. Process Dev.*, Vol. 19, No. 2, 1982, pp. 289–294.

$$VGC = \frac{SG - 0.24 - 0.022 \times \log(V_{210} - 35.5)}{0.755}$$

Where:

V = Saybolt Universal Viscosity @ 210°F in seconds

Refractive Index @ 20°C (68°F):

$$R_{i(20)} = \left(\frac{1 + 2 \times I}{1 - I}\right)^{1/2}$$

$$I = A \times \exp(B \times MeABP + C \times SG + D \times MeABP \times SG) \times MeABP^E \times SG^F$$

Constants

A	$2.341 * 10^{-2}$
B	6.464×10^{-4}
C	5.144
D	-3.289×10^{-4}
E	−0.407
F	−3.333

$$MW = a \times \exp(b \times MeABP + c \times SG + d \times MeABP \times SG) \times MeABP^e \times SG^f$$

Where:

Constants

a	20.486
b	1.165×10^{-4}
c	−7.787
d	1.1582×10^{-3}
e	1.26807
f	4.98308

Definitions of Fluidization Terms

Aeration. Any supplemental gas (air, steam, nitrogen, etc.) that increases fluidity of the catalyst.

Angle of Internal Friction—α. Angle of internal friction, or angle of shear, is the angle of solid against solid. It is the angle at which a catalyst will flow on itself in the nonfluidized state. For an FCC catalyst, this is about 80°.

Angle of Repose—β. The angle that the slope of a poured catalyst will make with the horizontal. For an FCC catalyst, this is typically 30°.

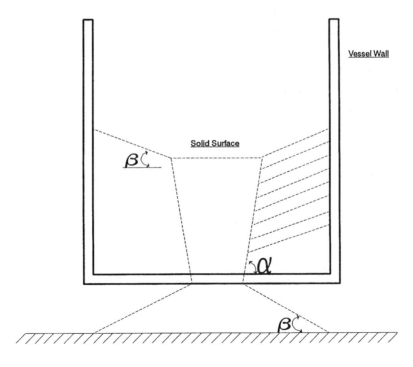

Apparent Bulk Density—ABD. The density of the catalyst at which it is shipped either in bulk volume or bags. It is density of the catalyst at minimum fluidization velocity.

Bed Density—ρ_b. The average density of a fluidized bed of solid particles and gas. Bed density is mainly a function of gas velocity and, to a lesser extent, the temperature.

Minimum Bubbling Velocity (Umb). The velocity at which discrete bubbles begin to form. Typical minimum bubbling velocity for an FCC catalyst is 0.03 ft/sec.

Minimum Fluidization Velocity (Umf). The lowest velocity at which the full weight of catalyst is supported by the fluidization gas. It is the minimum gas velocity at which a packed bed of solid particles will begin to expand and behave as a fluid. For an FCC catalyst, the minimum fluidization velocity is about 0.02 ft/sec.

Particle Density—ρ_p. The actual density of the solid particles taking into account any volume due to voids (pores) within the structure of the solid particles. Particle density is calculated as follows:

$$\rho_p = \frac{\text{Skeletal density}}{(\text{Skeletal density} \times \text{PV}) + 1}$$

Pore Volume—PV. The volume of pores or voids in the catalyst particles.

Ratio of Minimum Bubbling Velocity to Minimum Fluidization Velocity (Umb/Umf). This ratio can be calculated as follows:

$$\frac{\text{Umb}}{\text{Umf}} = \frac{2300 \times \rho_g^{0.126} \times \mu^{0.523} \times \exp^{0.716 \times F}}{d_p^{0.8} \times g^{0.934} \times (\rho_p - \rho_g)^{0.934}}$$

Where: ρ_g = gas density, kg/m^3
$\quad\quad\quad$ μ = gas viscosity, kg/m/sec
$\quad\quad\quad$ F = fraction of fines less than 45 microns
$\quad\quad\quad$ dp = mean particle size
$\quad\quad\quad$ ρ_p = particle density, kg/m^3
$\quad\quad\quad$ g = gravitational constant = 9.81 m/sec^2

The higher the ratio, the easier it is to fluidize the catalyst.

Skeletal Density—SD. The actual density of the pure solid materials that make up the individual catalyst particles. For an FCC catalyst, the skeletal density can be calculated as follows:

$$SD = \frac{100}{\dfrac{Al}{3.4} + \dfrac{Si}{2.1}}$$

Where: Al = Alumina content of the catalyst, wt%
Si = Silica content of the catalyst, wt%

Slip Factor. The ratio of vapor velocity to catalyst velocity.

Stick Slip Flow. The continuous sudden stoppage and resumption of catalyst flow in a standpipe. This is usually caused by underaeration.

Superficial Velocity. The velocity of the gas through the vessel or pipe without any solids present. It is a volumetric flow rate of fluidization gas divided by the cross-sectional area.

Conversion of ASTM 50% Point to TBP 50% Point Temperature

The following equation can be used to convert an ASTM D-86 50% temperature to a TBP 50% temperature.

$$TBP\ (50) = 0.87180 \times ASTM\ D\text{-}86\ (50)^{1.0258}$$

Where:

TBP(50) = true boiling point distillation temperature at 50 vol% distilled, °F

ASTM D86(50) = observed ASTM D-86 distillation temperature at 50 vol% distilled, °F

Example:

Given ASTM D-86(50) = 547°F, determine TBP 50% temperature:

$$TBP(50) = 0.87180 \times (547)^{1.0258}$$
$$TBP(50) = 561°F$$

Source: Daubert, T. E., "Petroleum Fraction Distillation Interconversions," *Hydrocarbon Processing,* September 1994, pp. 75–78.

Determination of TBP Cut Points from ASTM D-86

The difference between adjacent TBP cut points can be determined by the following equation:

$$Y_i = A \, X_i^B$$

Where:

Y_i = difference in TBP distillation between two cut points, °F
X_i = observed difference in ASTM D-86 distillation between two cut points, °F
A,B = constants varying for cut points ranges, shown in the following table:

i	Cut Point Range	A	B
1	100%–90%	0.11798	1.6606
2	90%–70%	3.0419	0.75497
3	70%–50%	2.5282	0.820072
4	50%–30%	3.0305	0.80076
5	30%–10%	4.9004	0.71644
6	10%–0%	7.4012	0.60244

Source: Daubert, T. E., "Petroleum Fraction Distillation Interconversions," *Hydrocarbon Processing,* September 1994, pp. 75–78.

$$\text{TBP (0)} \quad = \text{TBP(50)} - Y_4 - Y_5 - Y_6$$
$$\text{TBP (10)} = \text{TBP(50)} - Y_4 - Y_5$$
$$\text{TBP (30)} = \text{TBP(50)} - Y_4$$
$$\text{TBP (70)} = \text{TBP(50)} + Y_3$$
$$\text{TBP (90)} = \text{TBP(50)} + Y_3 + Y_2$$
$$\text{TBP (100)} = \text{TBP(50)} + Y_3 + Y_2 + Y_1$$

Conversion Factors

1 atmosphere (atm)	= 14.696 lb (force)/in^2, (absolute)
1 atmosphere (atm)	= 1.013 × 10^5 Newton/square meter (N/m^2)
1 atmosphere (atm)	= 1.013 bar
1 bar	= 10^5 pascal
1 barrel(bbl), 42 U.S.gal.	= 0.159 cubic meter (m^3)
1 barrel/day	= 6.625 × 10^{-3} m^3/hr
1 Btu	= 1,055 joule (J)
1 Btu	= 252.0 calories (cal)
1 Btu/hr	= 3.93 × 10^{-4} horsepower (hp)
1 Btu/hr	= 0.252 kcal/hr
1 Btu/hr	= 0.29307 Watts
1 Btu/lb	= 0.556 calorie/gram (cal/g)
1 Btu/lb	= 2.326 joules/gram (J/g)
1 Btu/lb—°F	= 4.186 joules/gram—°C
1 Btu/lb—°F	= 1.0 calorie/gram—°C
1 Btu/hr-ft^2—°F	= 4.882 kg-cal/hr-m^2—°C
degree Fahrenheit (°F)	= 1.8 × °C + 32
degree Kelvin (°K)	= °C + 273
degree Rankine (°R)	= 460 + °F
1 foot (ft or ')	= 12 inches (in or ")
1 foot (ft or ')	= 0.3048 meter (m)
1 gallon (gal), U.S.	= 3.785 × liters
1 gallon (gal), U.S.	= 3.785 × 10^{-3} cubic meter (m^3)
gas constant (°R)	= 10.73 (psia) × (ft^3)/(lb-mole) × (°R)
gas constant (°R)	= 8314 N/m^2 × m^3/kg-mole × °K
1 horsepower (hp)	= 746 watts (W)
1 inch (in. or ")	= 2.54 centimeters (cm)
1 inch (in. or ")	= 0.0254 meter (m)
1 pound (lb), weight	= 453.6 grams (g)
1 lb/ft^2-sec	= 4.8761 kg/m^2-sec
1 lb/ft^3	= 0.016 gram/cubic centimeter (g/cm^3)
1 lb/ft^3	= 0.016 gram/milliliter (g/ml)
1 lb/ft^3	= 16.018 kilogram/cubic meter (kg/m^3)

1 lb /gal (U.S.)	= 0.1198 g/cm^3
1 lb (force)/in^2 (psi)	= 0.0689 bars
1 lb (force)/in^2 (psi)	= 0.0680 atmospheres (atm)
1 lb (force)/in^2 (psi)	= 0.0703 kg/cm^2
1 mile	= 1.61 kilometers
1 ton (short)	= 2,000 pounds (lbs)
1 ton (short)	= 907.2 kilograms
1 ton (metric)	= 1000.00 kilograms
1 ton (long)	= 1016.0 kilograms
1 ton (long)	= 2,240 lbs

Glossary

Advance Process Control (APC) is a mechanism which manipulates regulatory controls toward more optimum unit operation.

Afterburn is the combustion of carbon monoxide (CO) to carbon dioxide (CO_2) in the dilute phase or in the cyclones of the regenerator.

Alkylation is one of the refining processes in which light olefin molecules are reacted with isobutane (in the presence of either sulfuric or hydrofluoric acid) to produce a "desirable" gasoline component called alkylate.

American Society of Testing and Materials (ASTM) is the organization that develops analytical tests and procedures to facilitate commerce.

Aniline Point is the minimum temperature for complete miscibility of equal volumes of aniline and the hydrocarbon sample. In cat cracking, aniline solution is used to determine aromaticity of FCC feedstocks. Aromaticity increases with reducing aniline point.

Anti-dumping is a federal requirement to prevent refiners from blending the "good" gasoline components in "reformulated gasoline" and the "bad" components in conventional gasoline.

Antimony is a metal, in either hydrocarbon or aqueous solution, commonly injected into the fresh feed to passivate nickel.

°API Gravity is an "artificial" scale of liquid gravity defined by: (141.5/SG) − 131.5. The scale was developed for water = 10. The main advantage of using °API gravity is that it magnifies small changes in liquid density.

Apparent Bulk Density (ABD) is the density of catalyst as measured, "loosely compacted" in a specified container.

Average Particle Size (APS) is the weighted average diameter of a catalyst.

Back-Mixing is the phenomena by which the catalyst travels more slowly up the riser than the hydrocarbon vapors.

Basic Nitrogen is the nitrogen compounds in the FCC feed that react with the catalyst acid sites, thereby reducing the catalyst's activity and selectivity.

Beta-Scission is splitting of the C–C bond two bonds away from the positively charged carbon atom.

Binder is the material used in the FCC catalyst to bind the matrix and zeolite components into a single homogeneous particle.

California Air Resources Board (CARB) is a state agency which regulates and sets standards for air quality and emissions of various pollutants.

Catalyst Activity is the conversion of feed (gas oils) to gasoline, lighter products, and coke in the MAT laboratory.

Carbenium Ion is a positively charged (R-CH2+) ion that is formed from a positive charge to an olefin and/or by removing a hydrogen and two electrons from a paraffin molecule.

Carbocation is a generic term for a positively charged carbon ion. Carbocation is further subdivided into carbenium and carbonium ions.

Carbonium Ion is a positively charged (CH5+) ion which is formed by adding a hydrogen ion (H+) to a paraffin.

Cat/Oil Ratio is the weight ratio of regenerated catalyst to the fresh feed in the riser feed injection zone.

Catalyst Cooler is a heat exchanger that removes heat from the regenerator through steam generation.

Cetane Number is a numerical indication of a fuel's (kerosene, diesel, heating oil) ignition quality. Cetane number is measured in a single-cylinder engine, whereas cetane index is a calculated value.

Coke is a hydrogen deficient residue left on the catalyst as a by-product of catalytic reactions.

Coke Factor is coke-forming characteristics of the equilibrium catalyst relative to coke-forming characteristics of a standard catalyst at the same conversion.

Coke (Carbon) on Regenerated Catalyst (CRC) is the level of residual of carbon remaining on the catalyst when the catalyst exits the regenerator.

Coke Yield is the amount of coke the unit produces to stay in heat balance, usually expressed as percent of feed.

Conradson Carbon, or Concarbon, is a standard test to determine the level of carbon residue present in a heavy oil feed.

Conventional Gasoline is a non-RFG gasoline that meets exhaust benzene, sulfur, olefins and T90 specifications.

Conversion is often defined as the percentage of fresh feed cracked to gasoline, lighter products, and coke. Raw conversion is calculated by subtracting the volume or weight percent of the FCC products (based on fresh feed) heavier than gasoline from 100, or:

Conversion = 100 − (LCO + HCO + DO) vol% or wt%

Cyclone is a centrifugal separator which collects and removes particulates from gases.

D-86 is a common ASTM test method that measures the boiling point of "light" liquid hydrocarbons at various volume percent fractions. The sample is distilled at atmospheric pressure, provided its final boiling point (end point) is less than 750°F.

D-1160 is an ASTM method that measures the boiling point of "heavy" liquid hydrocarbons at various volume percent fractions. The sample is distilled under vacuum (results are converted to atmospheric pressure). The application of D1160 is limited to a maximum final boiling point of about 1000°F.

Decanted Oil, Slurry, Clarified Oil, or Bottoms is the heaviest and often the lowest priced liquid product from a cat cracker.

Delta Coke is the difference between the coke content of the spent catalyst and the coke content of the regenerated catalyst. Numerical value of delta coke is calculated from:

Delta coke = coke yield (wt%/catalyst-to-oil ratio)

Dense Phase is the region where the bulk of the fluidized catalyst is maintained.

Dilute Phase is the region above the dense phase which has a substantially lower catalyst concentration.

Dipleg is the part of a cyclone separator that provides a barometric seal between the cyclone inlet and the cyclone solid outlet.

Distributive Control System (DCS) is a digital control system that has a distributive architecture where different control functions are implemented in specialized controllers.

Dynamic Activity is an indication of conversion per unit coke using data from the MAT laboratory.

Equilibrium Catalyst (E-cat) is the regenerated catalyst circulating from the reactor to the regenerator.

Exhaust Benzene is the amount of benzene toxins released. Exhaust benzene is a function of aromatics and benzene.

Expansion Joint is a mechanical assembly designed to eliminate large thermal stresses in the piping.

Faujasite is a naturally occurring mineral, having a specific crystalline, alumina-silicate structure, used in the manufacturing of the FCC catalyst. Zeolite faujasite is a synthetic form of the mineral.

Filler is the inactive component of the FCC catalyst.

Flapper Valve, Trickle Valve, or Check Valve is often attached to the end of a dipleg to minimize gas leakage up the dipleg and also catalyst losses during the unit start-up.

Free Radical is an uncharged molecule formed in the initial step of thermal cracking. Free radicals are very reactive and short-lived.

Gas Factor is the hydrogen and lighter gas-producing (C1-C4) characteristics of the equilibrium catalyst relative to the hydrogen and lighter gas-producing characteristics of some standard catalyst at the same conversion.

Heat of Cracking is the amount of energy required to convert FCC feed to the desired products.

Hydrogen Transfer is the secondary reaction that converts olefins (predominantly iso-olefins) into paraffins while extracting hydrogen from larger, more hydrogen-deficient molecules.

K Factor is an index designed to balance density and boiling point such that it relates solely to hydrogen content of the hydrocarbon.

Microactivity Test (MAT) is a small, packed-bed catalytic cracking test that measures activity and selectivity of a feedstock-catalyst combination.

Matrix is a substrate in which the zeolite is imbedded in the cracking catalyst. Matrix is often used as a term for the active, non-zeolitic component of the FCC catalyst.

Methyl Tertiary Butyl Ether (MTBE) is an ether added to gasoline to improve its octane and reduce air pollution.

Mix Zone Temperature is the theoretical equilibrium temperature between the regenerated catalyst and the uncracked vaporized feed at the bottom of the riser.

Molecular Sieve is a term applied to zeolite. Zeolite exhibits shape selectivity and hydrocarbon absorptions.

Motor Octane Number (MON) is a quantitative measure of a fuel to "knocking," simulating the fuel's performance under severe operating conditions (at 900 rpm and at 300°F).

n-d-M is an ASTM method that estimates chemical composition of a liquid stream.

Octane Barrel Yield, as used in the FCC, is defined as (RON + MON)/2 times the gasoline yield.

Oxygenate is an oxygen-containing hydrocarbon. The term is used for oxygen-containing molecules blended into gasoline to improve its combustion characteristics.

Particle Density is the actual density of solid particles, taking into account volume due to any voids (pores) within the structure of the solid particles.

Particle Size Distribution (PSD) is the particle size fractions of the FCC catalyst expressed as percent through a given sized hole.

Plenum is a means of collecting gases from multiple sets of cyclones before they are exhausted from the unit.

Pore Diameter is an estimate of the average pore size of the catalyst.

Pore Volume is the open space in the FCC catalyst, generally measured by mercury, nitrogen, or water. Mercury is used to measure large pores, nitrogen measures small pores, and water is used for both.

Ramsbottom, similar to Conradson Carbon, is a quantitative indication of carbon residue of a sample.

Rare Earth is a generic name used for the 14 metallic elements of the lanthanide series used in the manufacturing of FCC catalyst to improve stability, activity, and gasoline selectivity of the zeolite.

Reformulated Gasoline (RFG) is the gasoline sold in some ozone nonattainment metropolitan areas designed to reduce ozone and other air pollutants.

Refractive Index, similar to aniline point, is a quantitative indication of a sample's aromaticity.

Refractory is a cement-like material used to stand abrasion and erosion.

Reid Vapor Pressure (RVP) is gasoline vapor pressure at 100°F.

Resid refers to a process, such as resid cat cracking, that upgrades residual oil.

Residue is the residual material from the processing of raw crude (for example, vacuum *residue* and not vacuum resid).

Research Octane Number (RON) is a quantitative measure of a fuel to "knocking," simulating the fuel's performance under low engine severity (at 600 rpm and 120°F).

Riser is a vertical "pipe" where virtually all FCC reactions take place.

Selectivity is the ratio of yield to conversion for the "desired" products.

Silica Oxide to Alumina Oxide Ratio (SAR) is used to describe the framework composition of zeolite.

Skeletal Density is the actual density of the pure solid materials that make up individual particles.

Slide Valve or Plug Valve is a valve used to regulate the flow of catalyst between reactor and regenerator.

Slip Factor is the ratio of catalyst residence time to hydrocarbon vapors residence time in the riser.

Soda Y Zeolite is a "crystallized" form of Y-faujasite before any ion exchanges occur.

Spent Catalyst is the coke-laden catalyst in the stripper.

Standpipe is a means of conveying the catalyst between reactor and regenerator.

Stick-Slip Flow is erratic circulation caused when the catalyst packs and bridges across the standpipe.

True Boiling Point (TBP) is the distillation separation which has characteristics of 15 different theoretical plates at 5 to 1 reflux ratio.

Ultra Stable Y is a hydrothermally treated Y-faujasite which has a unit cell size at or below 24.50 °A and exhibits superior hydrothermal stability over Soda Y faujasite.

Unit Cell Size (UCS) is an indirect measure of active sites and SAR in the zeolite.

Zeolite is a synthetic crystalline alumina-silicate material used in the manufacturing of FCC catalyst.

Index